chance

**The science and secrets of
luck, randomness and
probability**

NewScientist

chance

The science and secrets of luck, randomness and probability

edited by Michael Brooks

JOHN MURRAY

First published in Great Britain in 2015 by Profile Books Ltd

This edition published in 2021 by John Murray (Publishers)
An Hachette UK company

2

Text design by Sue Lamble
Original New Scientist illustrations redrawn by Sue Lamble

A CIP catalogue record for this title is
available from the British Library

ISBN 978-1-473-64264-5
eBook ISBN 978-1-473-64265-2

Printed and bound in Great Britain by Clays Ltd, Elcograf S.p.A.

John Murray policy is to use papers that are natural,
renewable and recyclable products and made from
wood grown in sustainable forests. The logging and
manufacturing processes are expected to conform to the
environmental regulations of the country of origin.

John Murray (Publishers)
Carmelite House
50 Victoria Embankment
London EC4Y 0DZ

www.johnmurraypress.co.uk

Contents

Introduction

In 1989, a teenager called Richard Hill travelled north to Manchester, England, where he stayed with a friend of a friend. The next day, Ann, the friend's friend's mother, happened to be heading to Oxford. She offered Richard a lift south. He accepted.

During the journey, Richard mentioned that he lived in a nearby town called Swindon. 'Ah,' said Ann, 'maybe you know someone called Michael Brooks? He lives in Swindon. He'd be about twenty.'

There was a pause. 'He's engaged to my sister,' Richard said.

'Oh,' said Ann. 'He's my step-son.'

I had not seen Ann's husband – my father – since I was about a year old. But by chance, my future brother-in-law had stayed in my father's house. By chance, my father's wife had been heading in the same direction the next day. By chance the conversation had taken a turn that revealed a spooky connection between them.

I'm sure you have one of these stories. They defy explanation, and we can't help but imbue them with a deep significance. Richard, Ann and I – we are in regular contact thanks to that chance event – still don't know what to make of that extraordinary coincidence. It resonates with us as a kind of fundamental pivot point in all our lives. But should it?

To answer that, you have to understand what 'chance' actually is. And it turns out that this is much harder than you might expect.

'What are the chances?' It is a question that rings out every day, everywhere that humans exist. We don't usually have an answer – at least not one that's correct. Take the calculation by author Ali Binazir, who claimed, via a chain of reasoning about your mother and father meeting, eggs getting fertilised, and human longevity, that the odds of you existing are 1 in $10^{2,685,000}$ – a 10 followed by 2,685,000 zeroes.

Such odds are, at first glance, impressive. They create a sense of awe. But they are also nonsense. You are the result of all those things actually happening, whatever the odds of two random people falling in love, or a particular sperm fertilising a particular egg. And so is everyone else on the planet. There is no pool of people that didn't get born, so there is no way to calculate a probability of you existing. I hate to say it, but you're not, as Binazir claims, a miracle. You're just a link in the human chain.

Not that we can deny the role of chance in our universe. After all, it appears to be the most fundamental process in the laws of physics. Dig deep into the way everything works, and you find yourself dealing with quantum theory. This describes the world of the extremely small things from which all matter is made. Atoms, electrons, protons (and the quarks that are a proton's constituent parts) all obey the laws of quantum physics. And these laws are, in many ways, 'lawless'. There is no cause and effect at the heart of quantum theory: if I measure a property such as the spin of an electron it might be clockwise or anticlockwise. But the actual result of any single measurement is entirely unknowable in advance: it manifests at random. One of the most famous quotes in science is Einstein's reaction to this, a refusal to believe it can really be how the universe operates. 'God does not play dice,' he said to the physicist Niels Bohr.

Bohr's response was brilliant: he scolded Einstein for telling God what to do. He was right: our natural intuition that all effects must have a cause is not to be trusted. It evolved over millennia thanks to a need to survive in hostile landscapes. Our ancestors were better off assuming the bush over there is moving because of a tiger waiting to pounce than blithely assuming there is no reason for the rustling leaves. Evasive action may not always be strictly necessary, but it's the ultimate example of better safe than sorry.

For the same reasons, chance disarms us, causing us to attach significance to events that have no significance. Naively, we marvel at the discovery that two people at a party share the same birthday – another 'what are the chances?' But, if there are 23 or more people in the room, a shared birthday is statistically likely.

It's worth issuing a warning here: playing the birthday statistician is more likely to make you the party-pooper than the event's life and soul. That's because dealing properly with chance takes real mental effort, and parties aren't always the best place to demand that. However, chance is not just about gritty thinking; it can be a gateway to great fun and even unexpected success.

Understand the way human brains process chance and you could become the next world champion at rock-paper-scissors. Get to grips with its mathematical laws and you can make money betting on a football match – no matter who eventually wins the game. You might even be able to walk into a casino and beat the house – for a while, at least. Delve into those myths about people being naturally lucky, or star-cross'd and thus fated to suffer, and you'll discover that you can make your own luck.

Shakespeare's Romeo, the original star-cross'd lover, claimed that he was 'fortune's fool', created to fall into the hands of fate. But scientists are not sitting down and waiting for fate to determine their worthiness for a Nobel Prize. Instead they are skewing the odds in their favour by analysing serendipity and putting themselves in the best possible position to stumble across new discoveries. Louis Pasteur's contention that 'chance favours only the prepared mind' is one to take seriously, as it turns out.

Perhaps nowhere is the application of chance more serious than in the courtroom. If you have ever served on a jury, you'll have had the uncomfortable experience of making a life-changing decision (with the small comfort that it's always about someone else's life) based on much less information than you'd like. Rare is the open-and-shut case; instead, the jury's verdict hangs on its members' judgements of likelihood and probability. Here, again, our primitive brains often let us down. Even expert witnesses get chance wrong sometimes; no wonder moves are afoot to change the way we deal with chance in the legal sphere.

Within these pages you'll find these and many other revolutions-in-progress. Ways to rebel against the strictures of the digital world and put a little spice and unpredictability back into your life, for instance. You'll learn how to use surprise as a weapon, and even how best to find your lost car keys. Along the way you might have to confront the question of free will – do you have it? – and whether the future of the universe is fixed or yet to be written. But you will emerge with an understanding of the accidents that made you who you are, right back to the first moments after the big bang.

Chance is everywhere, and always has been. It was in the primordial quantum fluctuations that led to the formation of the Milky Way. It sparked the random genetic mutation that gave the first human brains access to unprecedented supplies of thought-fuelling glucose. It may even have played a role in putting this book into your hands today. Perhaps a friend or a lover bought it for you on a whim. Perhaps you bought it because you had just missed a train, wandered into the railway station's waiting room, and found a discarded newspaper that happened to contain a favourable review. Maybe you stumbled across it while browsing in a bookshop or library. It doesn't really matter; the important thing is that you grab this opportunity and read on. You are about to sit down to a banquet of brain food, and there's a possibility that this could be one of those unsought, serendipitous pivotal moments that changes everything. If chance really does favour the prepared mind, you're in luck.

Michael Brooks

Lucky to be here

Chance events from the big bang to the birth of humans

We are going to start our exploration of chance by tracing the chance events that led from the formation of the universe right up to the creation of human beings like you. Of course, there is no one exactly like you. Have you ever looked at a sibling and wondered where the differences come from? You might share the same genetic origins, but you aren't perfectly identical – not even if you are twins. Random twists and turns made you unique in the universe. The same seems to be true of the evolution of human life. It's an extraordinary journey, full of startling flukes. The universe didn't have to produce matter, or a planet with a stable enough climate for life to evolve. What's more, life – especially complex life – didn't have to evolve. Neither did species. By the time we get to the chance mutations that made humans what they are, you might just marvel at how lucky you are to exist.

Cosmic lottery

Let's begin at the beginning, where Stephen Battersby and David Shiga are on hand to explain our cosmological luck. This universe, it turns out, is something of a fluke.

What cosmic coincidences preceded our universe's birth are in the realms of speculation. Suffice it to say that some 13.82 billion years ago – give or take a yoctosecond – the cosmos was deciding what to be when it grew up.

'Much bigger', if the most popular model of the universe's beginnings is to be believed. According to the theory of inflation, the newly born universe was suffused with something called the inflaton field, which drove an exponential expansion of the cosmos for a period of about 10^{-32} seconds, stretching it flat and uniform in the process.

That usefully accounts for some otherwise tricky-to-explain characteristics of our universe, but the real point of interest is that the inflaton field, although essentially uniform, was not quite identical in each bit of space. Chance quantum fluctuations are responsible for this: they made it slightly more dense here, and just a little less dense there. It is lucky for us that this was so: perfect uniformity would have left the universe a very different place, uninteresting and almost certainly lifeless. As it is, one of those random microscopic quanta of noise, amplified by gravity, eventually grew into a huge agglomeration of galaxies and clusters of galaxies known as the Virgo supercluster. Among its many clumps is one straggly and undistinguished thicket we call the local group. Within that is the Milky Way, our home.

We know this thanks to the fact that, looking far out into the cosmos, astronomers can see the mottled pattern of the cosmic microwave background radiation. This is a snapshot of the growth and consolidation process in action when the first stable atoms formed some 380,000 years after the big bang. Variations in the pattern seem entirely random, and most physicists regard the quantum fluctuations that created it as having no cause at all. Of all happy accidents, this one might be the most accidental.

Then along came matter. It seems an extraordinary accident that matter exists at all: the cosmos could easily have been a bland sea of radiation. That's because, after inflation, the universe was still unimaginably hot and dense. It was filled with matter and antimatter particles – electrons, positrons, quarks, antiquarks and the like – buzzing around to no particular end. Stable unions between particles that might create stars, planets and life were still far away. And, most disturbing of all, matter and antimatter particles were present in what would have looked to the casual observer like equal numbers. That was a very precarious situation.

Standard theories say that matter and antimatter were created in equal amounts after the big bang. Since the two annihilate each other on contact, generating pairs of high-energy photons, all there should be in today's cosmos is vanilla-flavoured radiation. For us to exist, either matter or antimatter had to give – you can't make a planet or a person out of light.

Luckily, something seems to have favoured the creation of matter at a crucial moment within the first instants after the big bang. A surplus of just one extra matter particle per billion would have been enough to lead eventually to today's convenient matter-based residue. But how would such an imbalance arise?

While there is a pro-matter bias in some particle reactions, it is far too slight to create an advantage even this small. So physicists assume that a stronger bias, the consequence of unknown processes beyond the standard model of particle physics, must appear at the sort of high energies prevalent in the early universe.

An increasing suspicion is that such über-physics could be variable, changing across a multitude of universes. If so, then our little observable universe was fortunate to acquire a stock of matter, while many other realms will be radiation wastelands.

Matter is not the only potential victim of such shifty physics. It could also lead to some ultra-dense universes that collapse into black holes, and others laced with dark energy that rapidly rips all structures apart. In this picture, the emergence of a universe that would eventually become hospitable to humans is a rare prospect indeed.

The next cosmic accident was the advent of celestial fire. As we now appreciate, matter prevailed, and the universe cooled. Stable atoms and molecules soon formed. One hundred million years on, the first stars – giants of hydrogen and helium – appeared. They lived fast and died young in huge explosions that seeded the cosmos with heavier elements, the ingredients of later stars and galaxies. But a solar system doesn't form all by itself.

Until some 9 billion years after the big bang, hydrogen, helium and a sprinkling of the dust that fills the space between stars were all abundant in our corner of the cosmos. However, if they were to do anything more than hang around, something more was needed: a spark to set that inert gas cloud alight.

The spark did eventually come, and clues to its origin lie preserved in meteorites. Unlike the often-melted and mixed-up rocks native to our planet, meteorites have remained virtually unchanged since they condensed while the solar system was forming, preserving the chemistry of those early millennia.

One particular meteorite discovered in 2003 in Bishunpur, India, contained large quantities of iron-60, a radioactive isotope that decays over a few million years into stable nickel-60. Because iron-60 is so short-lived, interstellar gas generally holds just a trace of it. The large amounts in the Bishunpur meteorite imply that our solar system formed from a much richer brew.

There are two possible explanations. One is that the brew was spiced up by a nearby supernova. These massive stellar explosions are one of the few cosmic processes known to create large quantities of heavy radioactive isotopes such as iron-60. Shock waves from such a supernova could have triggered the formation of the sun and planets by compressing the primordial gas cloud.

The second possibility is that the conception of the solar system was a gentler affair. A red giant star of sufficient size could rival a supernova in iron-60 production and produce other radioactive elements in the right proportions to fit meteorite records. These elements would be forged in a deep layer of the star, carried to the surface by convection, and ejected as part of a powerful stellar wind that could also stir up any gas clouds nearby.

Astronomers think the supernova is the most likely explanation. But whether it was explosion or ejection, remember this: the sun is only the most obvious star we have to thank for our existence. Other, less famous ones should share the credit.

Next on our list of flukes is the formation of the moon. It was born from the fact that the solar system in which the infant Earth found itself was an unsettled environment, filled with lumps of rock whizzing around on irregular orbits. Some 4.5 billion years ago, one of these, something around the size of Mars, clobbered our planet. The result was a comprehensive rearrangement. Some of the impacting material stuck, while the rest was blasted into orbit along with bits of Earth excavated by the collision. There, it formed the moon.

It does not sound a particularly propitious event. But it was: it resulted in a satellite that is anomalously big in comparison to its parent planet. There is nothing else like it in the solar system, where all the other satellites are relatively small bodies that either accreted slowly from orbiting debris or were captured in passing.

It is similarly rare outside the solar system too. Observations by the Spitzer Space Telescope show us that giant collisions in other solar systems produce abundant dust. However, although a few such dusty systems have been found, collisions big enough to produce something like the moon seem to happen in only 5 to 10 per cent of solar systems. The number of instances where this has actually happened is considerably smaller even than that.

Why does this matter? Because the moon's size provides a steadying gravitational hand that helps to stabilise the tilt, or 'obliquity', of Earth's axis. That prevents wild changes in the pattern of solar heating on the planet's surface that could lead to extreme climate swings, including frequent periods where the whole planet freezes over. That's a big deal for us. If there were no moon, and obliquity varied significantly, the conditions for complex land-based life might not exist.

Life's origins owe a debt to another random astronomical event: pummelling by space rocks. It happened around 3.9 billion years ago, and is known as the late heavy bombardment. The cause of this impromptu game of planetary bagatelle is still not entirely clear. The best guess, though, is that it was triggered by a tussle between the solar system's four giant planets: Jupiter, Saturn, Uranus and Neptune. Slight drifts in Saturn and Jupiter's orbits eventually led to Saturn's orbital period becoming exactly twice that of Jupiter. This gravitational 'resonance' shook up the orbits of all four giant planets and sent nearby comets and asteroids shooting off towards the inner solar system.

The late heavy bombardment created extremely harsh conditions on Earth. 'Imagine pools of molten rock at the surface the size of the continent of Africa,' says Stephen Mojzsis, a geologist at the University of Colorado in Boulder. But once they had cooled, the impact craters would have been ideal sites in which to start life, says University of Edinburgh astrobiologist Charles Cockell, with residual heat driving chemical reactions in warm water circulating through the rock.

Alternatively, if life had already begun, the event would have altered the course of evolution, eliminating all but the most heat-tolerant microbes, says Mojzsis, adding 'This is the story of life – mass extinction leading to new styles of life.'

So, here we are, small beings on a small planet orbiting an unremarkable star in a really rather ordinary galaxy in an otherwise undistinguished part of an unimaginably vast universe. And how did we get started? By chance.

The algorithm of life

With all these accidents in place, Earth was ready for life. But that raises another question: did life have to happen? Paul Davies has spent much of his career pondering this question. The answer, he says, seems to lie in an unexpected field of study: computing.

Are we alone? The question of whether life is widespread in the universe is important. Researchers are searching for Earth-like planets around other stars – with some success – chiefly because they hope to find alien life there. Many assume that life should arise inevitably, given Earth-like conditions, a stance known as biological determinism. But it is hard to find any support for it in the known laws of physics, chemistry or biology. If we relied solely on these laws to explain the workings of the universe, it would be reasonable to conclude that life can only have arisen by sheer good luck – and that it is therefore exceedingly unlikely to be found elsewhere.

That said, those hoping to encounter aliens need not despair: research may yet justify the theory of biological determinism, and thereby boost our chances of finding neighbours somewhere in the cosmos.

This field received its first fillip in 1953, when Harold Urey and Stanley Miller at the University of Chicago tried to recreate in a test tube what they believed to be the conditions of primeval Earth. They found that amino acids – the building blocks of proteins – were part of the chemical sludge formed when electricity was discharged through a mixture of gaseous methane, ammonia, water vapour and hydrogen. The Miller–Urey experiment was hailed as the first step towards the creation of life in the laboratory: many chemists envisaged 'destination life' to lie at the end of a long road down which a chemical soup zapped with energy would be inexorably conveyed by the passage of time.

The idea quickly ran into problems, though. Making the building blocks of life is easy – amino acids have been found in meteorites and even in outer space. But just as bricks alone don't make a house, so it takes more than a random collection of amino acids to make life. Like house bricks, the building blocks of life have to be assembled in a very specific and exceedingly elaborate way before they have the desired function. To form proteins, many amino acids must link together in long chains in the right order. In energy terms that is an 'uphill' process.

In itself this is not a problem – there were plentiful energy sources on the early Earth. The problem is that, just as putting a stick of dynamite under a pile of bricks won't make a house, simply throwing energy willy-nilly at amino acids will not create delicate chain molecules with highly specific sequences. The likely result is simply a tarry mess.

Somehow the energy has to be fed into the system in a contrived and particular manner. In a living organism this step is under the control of the cell's molecular machinery, with its intricate specifications. In a jumbled prebiotic chemical soup, however, the amino acids would have to take pot luck. So while amino acids are written into the laws of nature, large and highly specialised molecules such as proteins are certainly not.

We now understand that the secret of life lies not with the basic chemical ingredients as such, but with the logical structure and organisational arrangements of the molecules. So DNA is a genetic databank, and genes are instructions for making customised proteins and, indirectly, other biological molecules. Like a supercomputer, life is an information-processing system, which implies a special sort of organised complexity. It is the information content, or software, of the living cell that is the real mystery, not the hardware components.

Nothing better illustrates the computational prowess of life than the genetic code. All known life is based on a deal struck between nucleic acids and proteins – two classes of molecule that, from a chemical point of view, are scarcely on nodding terms. The nucleic acids DNA and RNA store instructions, and proteins do most of the work. Together these molecules perform life's many miracles, but on their own they are helpless. To manufacture proteins, nucleic acids employ a clever intermediary to form a coded information channel.

It works like this. DNA, the famous double helix, is built like a ladder with four different kinds of rung. The information is stored in the sequences of these rungs, just as an instruction manual records information in sequences of letters. Proteins are built from 20 different amino acids, and the right protein is made only if the amino acids are linked together in the right order.

To translate from the four-letter alphabet used by DNA into the 20-letter system used by proteins, all known Earth life uses the same code. The key question when it comes to the inevitability – or otherwise – of life is how this ingenious system of coding emerged. How did stupid atoms spontaneously write their own software, and where did the very peculiar form of information needed to get the first living cell up and running come from?

Nobody knows the answer, but scientists have traditionally divided into two camps on the issue. In one group are those who believe it all happened by chance – that life is the result of a stupendous chemical fluke. It is easy to work out the odds against a random chemical mixture just happening to shuffle the appropriate molecules into the elaborate arrangement needed. The numbers are breathtakingly huge. If life as we know it arose by chance, it should have happened only once in the observable universe.

By contrast, biological determinists assume that chance is secondary, and that the right sorts of molecule obligingly form as a result of the laws of nature. American biogenesis pioneer Sidney Fox, for example, claimed that chemistry prefers to link up amino acids in precisely the right combinations to make them biologically functional. If so, it is as if there is an in-built bias – even a conspiracy – in nature to create life-encouraging substances. But is it credible that the laws of physics and chemistry contain a blueprint for life? How would the crucial information content of life be encoded in those laws?

To address this question, we need to think more carefully about the nature of the information that underpins living things. One important observation is that a structure that is rich in information tends to lack patterns. This property is illustrated most clearly by a branch of mathematics known as algorithmic information theory, which seeks to quantify the complexity of information by treating it as the output of a computer program, or algorithm.

Consider the binary sequence 10101010101010101010... This can be generated by the simple command 'Print 10 n times.' The input instructions are far shorter than the output sequence, reflecting the fact that the output contains a repeating pattern, which is easy to describe compactly. For this reason, the output has very little information content.

By contrast, an apparently random sequence such as 110101001010010111... cannot be condensed into a simple set of instructions, so it has a high information content. If the job of DNA is to store information efficiently, it had better not contain too many patterns in the sequence of 'rungs', since patterns represent informational redundancy. Biochemists confirm this expectation. The genomes of organisms that have been sequenced so far mostly look like random jumbles of the four constituent letters.

The higgledy-piggledy nature of genome sequences runs counter to biological determinism. The laws of physics can be used to predict ordered structures, but not random ones. A crystal, for instance, is simply a regular array of atoms with a periodic structure, like the repeating binary sequence given above, and is thus almost devoid of information. The construction of crystals is built into the laws of physics, as their periodic forms are determined by the mathematical symmetries inherent in those laws. But the random sequences of amino acids in proteins, or the series of 'rungs' in the DNA ladder, cannot be 'in' the laws of physics, any more than houses are.

Nor can they be 'in' the laws of chemistry. A direct illustration of this fact comes from examining the structure of DNA. Each rung of the ladder is made up of two segments, which couple together snugly like a lock and key. Ultimately, chemistry determines the nature of the bonds that hold together the segments, and also the forces that attach them to the sides of the ladder. However, there are no chemical bonds between successive rungs. Chemistry doesn't care about the order of the rungs, and life is free to change them on a whim. Just as the sequence of letters in an instruction manual is independent of the chemistry of the paper and ink, so the 'letters' in DNA – which make up the information – are independent of the chemical properties of nucleic acid. It is this ability of life to free itself from the strictures of chemistry that bestows upon it such power and versatility. Biological determinism would imply a chemical straitjacket that would serve only to inhibit, not enhance, biological creativity.

If life represents an escape from chemistry, we cannot appeal to chemistry to explain life. But where else might an explanation lie? Life is ultimately about complex information processing, so it makes sense to seek a solution in the realm of information theory and complexity theory.

Since biological information is not encoded in the laws

of physics and chemistry (at least as currently known), where does it come from? There seems to be agreement that information cannot come into existence spontaneously (except perhaps in the special case of the big bang), so the information content of living systems must somehow originate in their environment. Although there is no known law of physics able to create information from nothing, there might be some sort of principle that could explain how information can be garnered from the environment and accumulated in macromolecules.

One way to do this is by Darwinian evolution. Life on Earth started with simple organisms possessing short genomes with a relatively low information content. More complex organisms have longer genomes storing more information. The added information has flowed from the environment into the genomes by the process of natural selection: whenever a selection among alternative genomes is made – according to the degree of 'fitness' they confer on their owners – information is gained. So Darwinism can explain how organisms acquire information. But Darwinism kicks in only when life is already under way. How can we appeal to natural selection in the prebiotic phase?

Some biochemists believe that a form of molecular Darwinism is the answer. They envisage replicating molecules in some sort of chemical soup. Although bare replicating molecules may not satisfy most people's intuitive definition of life, they can still undergo a type of Darwinian evolution if they are subject to variation and selection. Proponents of this Darwinism-all-the-way-down theory suppose that the first replicator molecule was simple enough to form purely by chance.

The trouble is that the only experience we have of large replicating molecules is of those used by life. It is extremely unlikely that DNA would form by chance alone. Even its simpler cousin, RNA, is hard to make in long enough strands to be biologically potent. And shorter nucleic acid molecules tend to make more errors when replicating. If the error rate gets too high, information leaks away faster than selection can inject it, and evolution grinds to a halt. Far from accumulating information, an error-prone molecule will shed it.

So for molecular Darwinism to work, nature must obligingly provide replicators simple enough to form by chance, deft enough to replicate accurately and with a huge range of variants – which are also good replicators – for selection to act upon. These need not be nucleic acids, but to explain life as we know it they would eventually have to make nucleic acids and hand over the replicating function to them.

In effect, then, molecular Darwinism still smuggles in biological determinism. Not only must the laws of nature imply the existence of molecules possessing all the above properties, but the evolutionary pathway that the population of replicators follows must also lead to nucleic acids. Otherwise life as we know it would still be a tremendous fluke.

So should we concede that life is the result of an exceedingly unlikely chemical accident, a chance event unique in the entire universe? Not necessarily. A type of biological determinism may still be true, even if life isn't written into the familiar laws of physics, chemistry and evolutionary theory. It may be that the latter laws can account for life's hardware – that is, the raw materials – but the vital software, or informational component, derives from the laws of information theory.

The concept of 'information' is admittedly rather woolly, though this is usual when a subject is in its infancy. Two centuries ago, energy was an equally vague notion. Scientists intuitively recognised it as significant in physical processes, but it lacked mathematical rigour. Today, we accept energy as a real and fundamental quantity, because it is well understood. Information remains bewildering, partly because it crops up in different guises in so many scientific fields. In relativity theory, it is information that is forbidden to travel faster than light. In quantum mechanics, the state of a system is described by its maximum information content. In thermodynamics, information falls as entropy rises. In biology, a gene is a set of instructions containing the information needed to execute some task.

What we know about information comes mainly from the realm of human discourse. A landmark study in information theory was an analysis of communication over noisy radio channels conducted by American electrical engineer Claude Shannon during the Second World War. But nobody has yet written down the equivalent of Newton's laws for informational dynamics. Scientists can't even agree on whether information is invariably conserved in physical processes. For years, debate has raged over what happens to the information in a star when it collapses to form a black hole, which subsequently evaporates. Is the information irreversibly lost, or does it somehow get back out again?

One area of research, however, offers a tantalising pointer. Until relatively recently, biochemists treated life's molecules as little blocks that stick together. In reality, molecular structure and bonding are subject to quantum mechanics. Now physicists have extended the concept of information to the quantum domain, and made some extraordinary discoveries. One of these is the ability of quantum systems to process information exponentially faster than classical systems – a property that lies behind the quantum computer.

The riddle of biogenesis is essentially computational in nature. It involves discovering a very special type of molecular system from among a vast decision tree of chemical alternatives, most branches of which represent biological duds. Could it be that the key initial steps in 'informing' matter and setting it on the road to life lie in the offbeat realm of quantum physics? It remains an open question, but if the answer is yes, biological determinism might at last receive a convincing theoretical underpinning, justifying the popular belief that we inhabit a bio-friendly universe in which we are not alone.

A miraculous merger

Perhaps life is inevitable – but that is certainly not true of complex life. It is entirely conceivable that Earth's simplest life forms, which are little more than tiny bags of chemicals, could have remained unchanged. Your complex cells, with their internal compartments and complex support structures, their transportation fleets and their intricate machinery, might never have arisen. But then one day, 2 billion years ago, a fluke event occurred. The result? You, as Nick Lane explains.

We complex creatures are a rare and lucky breed. It would be surprising if simple cells like bacteria were not common throughout the universe. Organic molecules are formed from the reactions of the most ubiquitous of materials – water, rock and CO_2 – and they are thermodynamically close to inevitable. So the early appearance of simple bacterial cells on Earth, far from being a statistical quirk, is exactly what we would expect. However, if my work is right, complex life is not at all inevitable. It arose here just once in 4 billion years thanks to a rare, random event.

It all comes down to energy. Living things consume an extraordinary amount of energy, just to go on living. The food we eat gets turned into the fuel that powers all living cells, called ATP. This fuel is continually recycled: over the course of a day, humans each churn through 70 to 100 kilograms of the stuff. This huge quantity of fuel is made by enzymes, biological catalysts fine-tuned over aeons to extract every last drop of usable energy from reactions.

The enzymes that powered the first life cannot have been as efficient, and the first cells must have needed a lot more energy to grow and divide – probably thousands or millions of times as much energy as modern cells. The same must be true throughout the universe.

This phenomenal energy requirement is often left out of considerations of life's origin. What could the primordial energy source have been here on Earth? Old ideas of lightning or ultraviolet radiation just don't pass muster. Aside from the fact that no living cells obtain their energy this way, there is nothing to focus the energy in one place. The first life could not go looking for energy, so it must have arisen where energy was plentiful. In sunlight? Today, most life ultimately gets its energy from the sun, but photosynthesis is complex and probably didn't power the first life.

So what did? Reconstructing the history of life by comparing the genomes of simple cells is fraught with problems. Nevertheless, such studies all point in the same direction. The earliest cells seem to have gained their energy and carbon from the gases hydrogen and carbon dioxide. The reaction of H_2 with CO_2 produces organic molecules directly, and releases energy. That is important, because it is not enough to form simple molecules: it takes buckets of energy to join them up into the long chains that are the building blocks of life.

A second clue to how the first life got its energy comes from the energy-harvesting mechanism found in all known life forms. This mechanism was so unexpected that there were two decades of heated altercations after it was proposed by British biochemist Peter Mitchell in 1961.

Mitchell suggested that cells are powered not by chemical reactions, but by a kind of electricity. More specifically, by a difference in the concentration of protons, the charged nuclei of hydrogen atoms, across a membrane. Because protons have a positive charge, the concentration difference produces an electrical potential difference between the two sides of the membrane of about 150 millivolts. It might not sound like much, but because it operates over only 5 millionths of a millimetre, the field strength over that tiny distance is enormous, around 30 million volts per metre. That's equivalent to the field strength that creates a bolt of lightning.

Mitchell called this electrical driving force the proton-motive force. It sounds like a term from *Star Wars*, and that's not inappropriate. Essentially, all cells are powered by a force field as universal to life on Earth as the genetic code. This tremendous electrical potential can be tapped directly, to drive the motion of flagella, for instance, or harnessed to make the energy-rich fuel ATP.

The way in which life generates and taps this force field is extremely complex. The enzyme that makes ATP is a rotating motor powered by the inward flow of protons. Another protein that helps to generate the membrane potential, NADH dehydrogenase, works like a steam engine, with a moving piston for pumping out protons. These amazing nanoscopic machines must be the product of prolonged natural selection. They could not have powered life from the beginning. And that leaves us with a paradox.

Life guzzles energy, and inefficient primordial cells must have required much more energy, not less. These vast amounts of energy are most likely to have derived from a proton gradient, because the universality of this mechanism means it evolved early on. But how did early life manage something that today requires such sophisticated machinery?

There is a simple way to get huge amounts of energy this way. What's more, the context makes me think that it really wasn't that difficult for life to arise in the first place.

The answer I favour was proposed 25 years ago by the geologist Michael Russell, now at NASA's Jet Propulsion

Laboratory in Pasadena, California. Russell had been studying deep-sea hydrothermal vents. Say 'deep-sea vent' and many people think of dramatic black smokers surrounded by giant tube worms. Russell had something much more modest in mind: alkaline hydrothermal vents. These are not volcanic at all, and don't smoke. They are formed when seawater percolates down into the electron-dense rocks found in the Earth's mantle, such as the iron-magnesium mineral olivine.

Olivine and water react to form serpentinite in a process that expands and cracks the rock, allowing in more water and perpetuating the reaction. Serpentinisation produces alkaline – that is, proton-deficient – fluids rich in hydrogen gas. The heat this releases drives these fluids back up to the ocean floor. When they come into contact with cooler ocean waters, the minerals precipitate out, forming towering vents up to 60 metres tall.

Such vents, Russell realised, provide everything needed to incubate life. Or rather they did, 4 billion years ago. Back then, there was very little, if any, oxygen, so the oceans were rich in dissolved iron. There was probably a lot more CO_2 than there is today, which meant that the oceans were mildly acidic – that is, they had an excess of protons.

Just think what happens in a situation like this. Inside the porous vents, there are tiny, interconnected cell-like spaces enclosed by flimsy mineral walls. These walls contain the same catalysts – notably various iron, nickel and molybdenum sulphides – used by cells today (albeit embedded in proteins) to catalyse the conversion of CO_2 into organic molecules.

Fluids rich in hydrogen percolate through this labyrinth of catalytic micropores. Normally, it is hard to get CO_2 and H_2 to react: efforts to capture CO_2 to reduce global warming face exactly this problem. Catalysts alone may not be enough. But living cells don't capture carbon using catalysts alone – they use proton gradients to drive the reaction. And between a vent's alkaline fluids and acidic water there is a natural proton gradient.

Could this natural proton-motive force have driven the formation of organic molecules? It is too early to say for sure. I'm working on exactly that question, and there are exciting times ahead. But let's speculate for a moment that the answer is yes. What does that solve? A great deal. Once the barrier to the reaction between CO_2 and H_2 is down, the reaction can proceed apace. Remarkably, under conditions typical of alkaline hydrothermal vents, the combining of H_2 and CO_2 to produce the molecules found in living cells – amino acids, lipids, sugars and nucleobases – actually releases energy.

That means that, far from being some mysterious exception to the second law of thermodynamics, from this point of view life is in fact driven by it. It is an inevitable consequence of a planetary imbalance, in which electron-rich rocks are separated from electron-poor, acidic oceans by a thin crust, perforated by vent systems that focus this electrochemical driving force into cell-like systems. The planet can be seen as a giant battery; the cell is a tiny battery built on basically the same principles.

I'm the first to admit that there are many gaps to fill in, many steps between an electrochemical reactor that produces organic molecules and a living, breathing cell. But consider the bigger picture for a moment. The origin of life needs a very short shopping list: rock, water and CO_2.

Water and olivine are among the most abundant substances in the universe. Many planetary atmospheres in the solar system are rich in CO_2, suggesting that it is common too. Serpentinisation is a spontaneous reaction, and should happen on a large scale on any wet, rocky planet. From this perspective, the universe should be teeming with simple cells – life may indeed be inevitable whenever the conditions are right. It's hardly surprising that life on Earth seems to have begun almost as soon as it could.

Then what happens? It is generally assumed that once simple life has emerged, it gradually evolves into more complex forms, given the right conditions. But that's not what happened on Earth. After simple cells first appeared, there was an extraordinarily long delay – nearly half the lifetime of the planet – before complex ones evolved. What's more, simple cells gave rise to complex ones just once in 4 billion years of evolution: a shockingly rare anomaly, suggestive of a freak accident.

If simple cells had slowly evolved into more complex ones over billions of years, all kinds of intermediate cells would have existed and some still should. But there are none. Instead, there is a great gulf. On the one hand, there are the bacteria, tiny in both their cell volume and genome size: they are streamlined by selection, pared down to a minimum: fighter jets among cells. On the other, there are the vast and unwieldy eukaryotic cells, more like aircraft carriers than fighter jets. A typical single-celled eukaryote is about 15,000 times larger than a bacterium, with a genome to match.

All the complex life on Earth – animals, plants, fungi and so on – are eukaryotes, and they all evolved from the same ancestor. So without the one-off event that produced the ancestor of eukaryotic cells, there would have been no plants and fish, no dinosaurs and apes. Simple cells just don't have the right cellular architecture to evolve into more complex forms.

Why not? In 2010 I explored this issue with the pioneering cell biologist Bill Martin of the University of Düsseldorf in Germany. Drawing on data about the metabolic rates and genome sizes of various cells, we calculated how much energy would be available to simple cells as they grew bigger.

What we discovered is that there is an extraordinary energetic penalty for growing larger. If you were to expand a bacterium up to eukaryotic proportions, it would have tens of thousands of times less energy available per gene than an equivalent eukaryote. And cells need lots of energy per gene, because making a protein from a gene is an energy-intensive process. Most of a cell's energy goes into making proteins.

At first sight, the idea that bacteria have nothing to gain by growing larger would seem to be undermined by the fact that there are some giant bacteria bigger than many complex cells, notably *Epulopiscium*, which thrives in the gut of the surgeonfish. Yet *Epulopiscium* has up to 200,000 copies of its complete genome. Taking all these multiple genomes into consideration, the energy available for each copy of any gene is almost exactly the same as for normal bacteria, despite the vast total amount of DNA. They are perhaps best seen as consortia of cells that have fused together into one, rather than as giant cells.

So why do giant bacteria need so many copies of their genome? Recall that cells harvest energy from the force field across their membranes, and that this membrane potential equates to a bolt of lightning. Cells get it wrong at their peril. If they lose control of the membrane potential, they die. Nearly 20 years ago, biochemist John Allen, now a colleague at University College London, suggested that genomes are essential for controlling the membrane potential, because they can control protein production. These genomes need to be near the membrane they control so they can respond swiftly to local changes in conditions. Allen and others have amassed a good deal of evidence that this is true for eukaryotes, and there are good reasons to think it applies to simple cells too.

So the problem that simple cells face is this. To grow larger and more complex, they have to generate more energy. The only way they can do this is to expand the area of the membrane they use to harvest energy. To maintain control of the membrane potential as the area of the membrane expands, though, they have to make extra copies of their entire genome – which means they don't actually gain any energy per gene copy.

Put another way, the more genes that simple cells acquire, the less they can do with them. And a genome full of genes that can't be used is no advantage. This is a tremendous barrier to cells growing more complex, because making a fish or a tree requires thousands more genes than bacteria possess.

So how did eukaryotes get around this problem? By acquiring mitochondria.

About 2 billion years ago, one simple cell somehow ended up inside another. The identity of the host cell isn't clear, but we know it acquired a bacterium, which began to divide within it. These cells within cells competed for succession; those that replicated fastest, without losing their capacity to generate energy, were likely to be better represented in the next generation.

And so on, generation after generation, these endosymbiotic bacteria evolved into tiny power generators, containing both the membrane needed to make ATP and the genome needed to control membrane potential. Crucially, though, along the way they were stripped down to a bare minimum. Anything unnecessary has gone, in true bacterial style. Mitochondria originally had a genome of perhaps 3,000 genes; nowadays they have just 40 or so genes left.

For the host cell, it was a different matter. As the mitochondrial genome shrank, the amount of energy available per host-gene copy increased and its genome could expand. Awash in ATP, served by squadrons of mitochondria, it was free to accumulate DNA and grow larger. You can think of mitochondria as a fleet of helicopters that 'carry' the DNA in the nucleus of the cell. As mitochondrial genomes were stripped of their own unnecessary DNA, they became lighter and could each lift a heavier load, allowing the nuclear genome to grow ever larger.

These huge genomes provided the genetic raw material that led to the evolution of complex life. Mitochondria did not prescribe complexity, but they permitted it. It's hard to imagine any other way of getting around the energy problem – and we know it happened just once on

Earth because all eukaryotes descend from a common ancestor.

The emergence of complex life, then, seems to hinge on a single fluke event – the acquisition of one simple cell by another. Such associations may be common among complex cells, but they are extremely rare in simple ones. And the outcome was by no means certain: the two intimate partners went through a lot of difficult co-adaptation before their descendants could flourish.

This means there is no inevitable evolutionary trajectory from simple to complex life. Never-ending natural selection, operating on infinite populations of bacteria over billions of years, may never give rise to complexity. Bacteria simply do not have the right architecture. They are not energetically limited so long as they remain small in genome size and cell volume – the problem only becomes visible when we look at what it would take for their volume and genome size to expand. Only then can we see that bacteria occupy a deep canyon in an energy landscape, from which they are unable to escape.

This line of reasoning suggests that while Earth-like planets may teem with life, very few ever give rise to complex cells. That means there are very few opportunities for plants and animals to evolve, let alone intelligent life. So even if we were to discover that simple cells evolved on Mars, too, this wouldn't tell us much about how common animal life is elsewhere in the universe.

All this might help to explain why we've never found any sign of aliens. Of course, some of the other explanations that have been proposed, such as life on other planets usually being wiped out by catastrophic events such as gamma-ray bursts long before smart aliens get a chance to evolve, could well be true too. If so, there may be very few other intelligent aliens in the galaxy.

Then, again, perhaps some just happen to live in our neighbourhood. If we do ever meet them, there's one thing I would bet on: they will have mitochondria too.

The accident of species

One cell accidentally acquires another, and complex life is born. Even then, however, you are not an inevitable result. Earth is teeming with a multitude of animal and plant species, and this diversity is far more of an accident than biologists ever wanted to believe. Here's Bob Holmes.

Antarctic fish deploy antifreeze proteins to survive in cold water. Tasty viceroy butterflies escape predators by looking like toxic monarchs. Disease-causing bacteria become resistant to antibiotics. Everywhere you look in nature, you can see evidence of natural selection at work in the adaptation of species to their environment. Surprisingly, though, natural selection may have little role to play in one of the key steps of evolution – the origin of new species. Instead it would appear that speciation is merely an accident of fate.

So, at least, says Mark Pagel, an evolutionary biologist at the University of Reading, UK. If his controversial claim proves correct, then the broad canvas of life – the profusion of beetles and rodents, the dearth of primates, and so on – may have less to do with the guiding hand of natural selection and more to do with evolutionary accident-proneness.

Of course, there is no question that natural selection plays a key role in evolution. Darwin made a convincing case a century and a half ago in *On the Origin of Species*, and countless subsequent studies support his ideas. But there is an irony in Darwin's choice of title: his book did not explore what actually triggers the formation of a new species. Others have since grappled with the problem of how one species becomes two, and with the benefit of genetic insight, which Darwin lacked, you might think they would have cracked it. Not so. Speciation still remains one of the biggest mysteries in evolutionary biology.

Even defining terms is not straightforward. Most biologists see a species as a group of organisms that can breed among themselves but not with other groups. There are plenty of exceptions to that definition – as with almost everything in biology – but it works pretty well most of the time. In particular, it focuses attention on an important feature of speciation: for one species to become two, some subset of the original species must become unable to reproduce with its fellows.

How this happens is the real point of contention. By the middle of the 20th century, biologists had worked out that reproductive isolation sometimes occurs after a few organisms are carried to newly formed lakes or far-off islands. Other speciation events seem to result from major changes in chromosomes, which suddenly leave some individuals unable to mate successfully with their neighbours.

It seems unlikely, though, that such drastic changes

alone can account for all or even most new species, and that's where natural selection comes in. Species exist as more or less separate populations in different areas, and the idea here is that two populations may gradually drift apart, like old friends who no longer take the time to talk, as each adapts to a different set of local conditions. 'I think the unexamined view that most people have of speciation is this gradual accumulation by natural selection of a whole lot of changes, until you get a group of individuals that can no longer mate with their old population,' says Pagel.

For a long time, no one could find a way to test whether this hunch really does account for the bulk of speciation events, but more than a decade ago Pagel came up with an idea of how to solve this problem. If new species are the sum of a large number of small changes, he reasoned, then this should leave a telltale statistical footprint in their evolutionary lineage.

Whenever a large number of small factors combine to produce an outcome – whether it be a combination of nature and nurture determining an individual's height, economic forces setting stock prices, or the vagaries of weather dictating daily temperatures – a big enough sample of such outcomes tends to produce the familiar bell-shaped curve that statisticians call a normal distribution. For example, people's height varies widely, but most heights are clustered around the middle values. So, if speciation is the result of many small evolutionary changes, Pagel realised, then the time interval between successive speciation events – that is, the length of each branch in an evolutionary tree – should also fit a bell-shaped

On the origin of species

Which processes drive speciation? Chopping up an evolutionary tree and plotting the number of twigs of different lengths gives characteristic curves for different species

Phylogenetic tree representing speciation within a given group of organisms

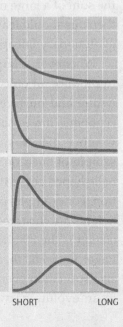

Branch points indicate a split of a single species

Length of twig represents longevity of a species before it splits

Exponential curve

Indicates that speciation is triggered by **a single accidental event**. Best fit for **78%** of evolutionary trees

Variable rate curve

Indicates that speciation happens in bursts when **environmental conditions offer new opportunities**. Just **6%** of trees fit this pattern

Bell curve (lognormal distribution)

Species arise gradually, with many changes multiplying until they reach a threshold. Accounts for **8%** of trees

Bell curve (normal distribution)

Species arise gradually, with many changes adding up until they reach a threshold. No trees conform to this pattern

SHORT   LONG

distribution (see diagram). That insight, straightforward as it was, ran into a roadblock, however: there simply weren't enough good evolutionary trees to get an accurate statistical measure of the branch lengths. So Pagel filed his idea away and got on with other things.

Then, a few years ago, he realised that reliable trees had suddenly become abundant, thanks to cheap and speedy DNA-sequencing technology. 'For the first time, we have a large tranche of really good phylogenetic trees to test the idea,' he says. So he and his colleagues Chris Venditti and Andrew Meade rolled up their sleeves and got stuck in.

The team gleaned more than 130 DNA-based evolutionary trees from the published literature, ranging widely across plants, animals and fungi. After winnowing the list to exclude those of questionable accuracy, they ended up with a list of 101 trees, including various cats, bumblebees, hawks, roses and the like.

Working with each tree separately, they measured the length between each successive speciation event, essentially chopping the tree into its component twigs at every fork. Then they counted up the number of twigs of each length, and looked to see what pattern this made. If speciation results from natural selection via many small changes, you would expect the branch lengths to fit a bell-shaped curve. This would take the form of either a normal curve if the incremental changes sum up to push the new species over some threshold of incompatibility, or the related lognormal curve if the changes multiply together, compounding one another to reach the threshold more quickly.

To their surprise, neither of these curves fitted the data.

The lognormal was best in only 8 per cent of cases, and the normal distribution failed resoundingly, providing the best explanation for not a single evolutionary tree. Instead, Pagel's team found that in 78 per cent of the trees, the best fit for the branch-length distribution was another familiar curve, known as the exponential distribution.

Like the bell curve, the exponential has a straightforward explanation – but it is a disquieting one for evolutionary biologists. The exponential is the pattern you get when you are waiting for some single, infrequent event to happen. The time interval between successive telephone calls you receive fits an exponential distribution. So does the length of time it takes a radioactive atom to decay, and the distance between roadkills on a highway.

To Pagel, the implications for speciation are clear: 'It isn't the accumulation of events that causes a speciation, it's single, rare events falling out of the sky, so to speak. Speciation becomes an arbitrary, happy accident when one of these events happens to you.'

All kinds of rare events could trigger the accident of speciation. Not just physical isolation and major genetic changes, but environmental, genetic and psychological incidents. The uplift of a mountain range that split a species in two could do it. So too could a mutation that made fish breed in surface waters instead of near the bottom, or a change in preference among female lizards for mates with blue spots rather than red ones.

The key point emerging from the statistical evidence, Pagel stresses, is that the trigger for speciation must be some single, sharp kick of fate that is, in an evolutionary sense, unpredictable.

'We're not saying that natural selection is wrong, that Darwin got it wrong,' Pagel adds. Once one species has split into two, natural selection will presumably adapt each to the particular conditions it experiences. The point is that this adaptation follows as a consequence of speciation, rather than contributing as a cause. 'I think what our paper points to – and it would be disingenuous for very many other people to say they had ever written about it – is what could be, quite frequently, the utter arbitrariness of speciation. It removes speciation from the gradual tug of natural selection drawing you into a new niche,' he says.

This has implications for one of the most contentious aspects of evolution: whether it is predictable or not. If Pagel is correct, natural selection shapes existing species in a gradual and somewhat predictable way, but the accidental nature of speciation means that the grand sweep of evolutionary change is unpredictable. In that sense his findings seem to fit with the famous metaphor of the late Stephen Jay Gould, who argued that if you were able to rewind history and replay the evolution of life on Earth, it would turn out differently every time.

Other evolutionary biologists have been reluctant to accept Pagel's idea wholeheartedly. Some regard it as interesting but in need of further testing. 'The single, rare events model is brilliant as an interpretation – as a potential interpretation,' is how Arne Mooers at Simon Fraser University in Vancouver, Canada, reacts. Others suspect Pagel's analysis has highlighted only part of the story. 'It's telling you about one necessary but not sufficient component of speciation,' says Daniel Rabosky of the University of Michigan. 'You have to have two things: something to cause isolation, and something to cause differentiation.' And the latter process – through which the two isolated populations change enough that we recognise them as two distinct species – is likely to involve gradual, adaptive change under the hand of natural selection.

The notion that the formation of a new species has little to do with adaptation sits uncomfortably with fundamental ideas about evolution. A particular stumbling block is what evolutionary biologists call 'adaptive radiations'. When ecological opportunities open up – as, for example, when finches from the South American mainland first colonised the Galapagos – species seem to respond by diversifying into a host of new forms, each adapted to a particular niche. These bursts of speciation suggest that organisms need not wait for some rare event to push them into speciating, but instead can be pulled there by natural selection.

In his analysis, Pagel specifically looked for the signature of this kind of evolutionary exuberance. Bursts of speciation would manifest as trees with lots of branching at irregular intervals; in other words, a highly variable rate of change over time, giving rise to a subtly different curve. 'It was the model that, going in, I thought would explain far and away the most trees,' says Pagel.

He was wrong. 'When it works, it works remarkably well,' he says. 'But it only works in about 6 per cent of cases. It doesn't seem to be a general way that groups of species fill out their niches.'

This finding has independent support. Luke Harmon at the University of Idaho in Moscow and his colleagues have examined 49 evolutionary trees to see whether there are bursts of evolutionary change early in a group's history, when unfilled niches might be expected to be most common. There is little evidence for such a pattern, they report in a paper published in the journal *Evolution*.

If speciation really is a happy accident, what does that mean for the way biologists study it? By focusing on the selective pressures that drive two species into different ecological niches, as they currently do, they may learn a lot about adaptation but not much about speciation. 'If you really want to understand why there are so many rodents and so few of other kinds of mammals, you should start to look at the catalogue of potential causes of speciation in an animal's environment, rather than take the view that there are all these niches out there that animals are constantly being drawn into,' Pagel says.

Rodents adapted to cool climates, for example, would be prone to isolation on mountain tops if the climate warmed. That could make them more likely to speciate than mammals adapted to warm temperatures. Likewise, marine animals whose larval stages live on the sea floor might be more likely to split into separate isolated populations and therefore speciate more often than those with free-floating larvae. Indeed, this is exactly what palae-ontologist David Jablonski of the University of Chicago has found among marine snails. Similarly, species with narrow habitat requirements or finicky mate-choice rituals may also be prone to accidental splits.

What other possible accidents might there be? No one knows yet. 'We'd like people to start compiling the lists of these things that might lead to speciation, and start making predictions about who's going to have a high rate of speciation and who's going to have a low rate,' says Pagel. If these lists help us understand the broad sweep of evolutionary history – the rise of mammals, why there are so many species of beetles, or the remarkable success of flowering plants – then we will know Pagel is onto something fundamental.

In the meantime, though, Pagel's take on speciation may help explain another puzzling feature of the natural world. Over and over again, as biologists sequence the DNA of wild organisms, they find that what appears superficially to be a single species is actually two, several or even many. The forests of Madagascar are home to 16 different species of mouse lemurs, for example, all of which live in similar habitats, do similar things, and even look pretty much alike. These cryptic species complexes are difficult to explain if speciation is the end result of natural selection causing gradual divergence into different niches. But if new species are happy accidents, there need be no ecological difference between them.

Pagel's own epiphany in this regard came in Tanzania, as he sat at the base of a hardwood tree watching two species of colobus monkey frolic in the canopy 40 metres overhead. 'Apart from the fact that one is black and white, and one is red, they do all the same things,' he says. 'I can remember thinking that speciation was very arbitrary. And here we are – that's what our models are telling us.'

Lucky you!

*From quantum fluctuations in the cosmic void, through bombard-
ments by comets and the accidental creation of Earth's myriad
species, we finally arrived at you. And even here, blind luck is
responsible for much of what makes humans special. Clare Wilson
investigates the final spin of our evolutionary wheel of fortune,
taking a look at the raft of chance mutations that created humans.*

Earth, several million years ago: a cosmic ray from space
blasts into the atmosphere at close to the speed of light.
It collides with an oxygen atom, generating a shower of
energetic particles, one of which knocks into a DNA molecule within a living creature.

That DNA molecule happens to reside in a developing egg cell within an ape-like animal living in Africa.
The DNA is altered by the collision – mutated – and the
resulting offspring is slightly different from its mother.
The mutation gives the offspring an advantage over its
peers in the competition for food and mates, and so, as
the generations pass, it is carried by more and more of
the population. Eventually it is almost ubiquitous, and so
the altered version of the DNA should really no longer be
called a mutation – it's just one of the regular 23,000 or so
genes that make up the human genome.

While cosmic rays are thought to be one source of
mutations, DNA-copying errors during egg and sperm
production may be a more common cause. Whatever
their origins, these evolutionary accidents took us on a
6-million-year journey from something similar to a great
ape to us, *Homo sapiens*.

It was a remarkable transformation, and yet we have only recently started to gain insight into the mutations that might have been involved. For a long time, most of our knowledge of human evolution had to be gleaned from fragments of bone found in the earth – a bit like trying to work out the picture on a jigsaw when most of the pieces are missing. The fraction of animal remains that happen to be buried under the right conditions to fossilise can only be guessed at, but it is likely to be vanishingly small.

That is why the field of palaeoanthropology has been given such a boost by the explosion in genetic-sequencing technologies. In 2003, a complete read-out of the human genome was published, a project that took 13 years. Since then, thanks to the technology getting faster and cheaper, barely a year goes by without another genome rolling off the production line. We have now sequenced creatures including chimpanzees, gorillas and orang-utans, as well as Neanderthals and Denisovans, our distant cousins who left Africa before *Homo sapiens* did. We are a million miles from a complete list, but even the first few to emerge as likely candidates are shedding light on the ascent of man. 'It gives us a perspective on what it takes to become human,' says John Hawks, a palaeoanthropologist at the University of Wisconsin–Madison.

Comparing these genomes reveals a wealth of information. If a gene that is active in the brain is different in humans and chimps, for instance, that could point to a mutation that helped to make us smarter. In fact, comparing the human and chimp genomes reveals about 15 million substitutions in the 'letters' that make up the genetic code. There are also wholesale deletions of DNA or duplications. Based on what we already know about DNA, the vast majority of these changes would not have affected our physical traits. That's either because the change to the DNA is so minor that it would not influence a gene's function, or because the mutation is in a region of 'junk' DNA. It is estimated that out of the 15 million differences, perhaps 10,000 were changes to genes that altered our bodies and were therefore subject to natural selection.

It's still a formidable target, and that's not counting mutations to the regulatory regions of our DNA, which act as on/off switches for genes. It is not yet possible to calculate a figure for this type of mutation in the human line, although they are thought to have played a crucial role in evolution.

So far several hundred mutations have been identified that affected us. More discoveries will follow, but documenting the DNA changes is not half as challenging as working out what they did. 'Determining their effect requires immense experimentation and sometimes the creation of transgenic animals,' says Hawks. 'This is difficult science to undertake. We are at the very early stages.'

Even so, we have already had a glimpse of many of the pivotal points in human evolution, including the rapid expansion of our brains, the emergence of speech and the possible origin of our opposable thumbs. Read on to discover the six evolutionary accidents that made you the person you are today.

1) Jaw dropper

A chimpanzee's jaws are so powerful it can bite off a person's finger in one chomp. That is not a theoretical calculation; more than one primate researcher has lost a digit that way.

Humans have wimpy jaw muscles by comparison. This could be down to a single mutation in a gene called *MYH16*, which encodes a muscle protein. The mutation inactivates the gene, causing our jaw muscles to be made from a different version of the protein. They are consequently much smaller.

This finding, which came in 2004, caused a stir when the researchers argued that smaller jaw muscles could have allowed the growth of a bigger skull. Primates with big jaw muscles have thickened supporting bone at the back of their skull, which arguably constrains skull expansion, and therefore that of the brain too. 'We are suggesting this mutation is the cause of the decrease in muscle mass and hence the decrease in bone,' says Hansell Stedman, a muscle researcher at the University of Pennsylvania in Philadelphia, who led the work. 'Only then do you lift the evolutionary constraint that precludes other mutations that allow your brain to continue growing.'

The team dated the mutation to 2.4 million years ago – just before our brain expansion took off. But another study, which sequenced a longer section of the muscle gene, came up with an earlier estimate for when the mutation occurred – 5.3 million years ago.

Whichever date is right, the mutation still happened after we split from our last common ancestor with chimps.

Why would our ancestors switch to a weaker bite? Stedman speculates that rather than changes in diet being the catalyst, it could be that our ancestors no longer used biting as a form of attack. 'At some point, perhaps through social organisation, this form of weaponry became more optional for our ancestors,' he says.

2) Brain gain

Our braininess is one of our species' defining features. With a volume of 1,200 to 1,500 cubic centimetres, our brains are three times the size of those of our nearest relative, the chimpanzee. This expansion may have involved a kind of snowball effect, in which initial mutations caused changes that were not only beneficial in themselves but also allowed subsequent mutations that enhanced the brain still further. 'You have some changes and that opens opportunities for new changes that can help,' Hawks says.

In comparison with that of a chimp, the human brain has a hugely expanded cortex, the folded outermost layer that is home to our most sophisticated mental processes, such as planning, reasoning and language abilities. One approach to finding the genes involved in brain expansion has been to investigate the causes of primary microcephaly, a condition in which babies are born with a brain one-third of the normal size, with the cortex particularly undersized. People with microcephaly are usually cognitively impaired to varying degrees.

Genetic studies of families affected by primary microcephaly have so far turned up seven genes that can cause the condition when mutated. Intriguingly, all seven play a role in cell division, the process by which immature neurons multiply in the fetal brain, before migrating to their final location. In theory, if a single mutation popped up that caused immature neurons to undergo just one extra cycle of cell division, that could double the final size of the cortex.

Take the gene *ASPM*, short for 'abnormal spindle-like microcephaly-associated'. It encodes a protein found in immature neurons that is part of the spindle – a molecular scaffold that shares out the chromosomes during cell division. We know this gene was undergoing major changes just as our ancestors' brains were rapidly expanding. When the human *ASPM* sequence was compared with that of seven primates and six other mammals, it showed several hallmarks of rapid evolution since our ancestors split from chimpanzees.

Other insights come from comparing the human and chimp genomes to pin down which regions have been evolving the fastest. This process has highlighted a region called HAR1, short for human accelerated region-1, which is 118 DNA base pairs long. We do not yet know what HAR1 does, but we do know that it is switched on in the fetal brain between 7 and 19 weeks of gestation, in the cells that go on to form the cortex. 'It's all very tantalising,' says Katherine Pollard, a biostatistician at The Gladstone Institutes in San Francisco, who led the work.

Equally promising is the discovery of two duplications of a gene called *SRGAP2*, which affect the brain's development in the womb in two ways: the migration of neurons from their site of production to their final location is accelerated, and the neurons extrude more spines, which allow neural connections to form. According to Evan Eichler, a geneticist at the University of Washington in Seattle who was involved in the discovery, those changes 'could have allowed for radical changes in brain function'.

3) Energy upgrade

While it is tough to work out just how our brains got so big, one thing is certain: all that thinking requires extra energy. The brain uses about 20 per cent of our energy at rest, compared with about 8 per cent for other primates. 'It's a very metabolically demanding tissue,' says Greg Wray, an evolutionary biologist at Duke University in Durham, North Carolina.

Three mutations have been discovered that may have helped meet that demand. One emerged with the publication of the gorilla genome. This revealed a DNA region that underwent accelerated evolution in an ancient primate ancestor, common to humans, chimps and gorillas, some time between 15 and 10 million years ago.

The region was within a gene called *RNΓ213*, the site of a mutation that causes Moyamoya disease – a condition that involves a narrowing of the arteries to the brain. That suggests the gene may have played a role in boosting the brain's blood supply during our evolution. 'We know that damaging the gene can affect blood flow, so we can speculate that other changes might influence that in a beneficial way,' says Chris Tyler-Smith, an evolutionary geneticist at the Sanger Institute in Cambridge, UK, who was part of the group that sequenced the gorilla genome.

There are more ways to boost the brain's energy

supply than just replumbing its blood vessels, though. The organ's main food source is glucose and this is drawn into the brain by a glucose-transporter-molecule in the blood-vessel walls.

Compared with chimpanzees, orang-utans and macaques, humans have slightly different 'on switches' for two genes that encode the glucose transporters for brain and muscle, respectively. The mutations mean more glucose transporters in our brain capillaries and fewer in our muscle capillaries. 'It's throwing a switch so you divert a greater fraction [of the available glucose] into the brain,' says Wray. In short, it looks like athleticism has been sacrificed for intelligence.

4) Gift of the gab

Bring up a chimpanzee from birth as if it were a human and it will learn many unsimian behaviours, like wearing clothes and even eating with a knife and fork. But one thing it will not do is talk.

In fact, it would be physically impossible for a chimp to talk just like us, thanks to differences in our voice boxes and nasal cavities. There are neurological differences too, some of which are the result of changes to what has been dubbed the 'language gene'.

This story began with a British family that had 16 members over three generations with severe speech difficulties. Usually speech problems are part of a broad spectrum of learning difficulties, but the 'KE' family, as they came to be known, seemed to have deficits that were more specific. Their speech was unintelligible and they had a hard time understanding others' speech, particularly when it involved applying rules of grammar. They also had problems making complex movements of the mouth and tongue.

In 2001, the problem was pinned on a mutation in a gene called *FOXP2*. We can tell from its structure that the gene helps regulate the activity of other genes. Unfortunately, we do not yet know which ones are controlled by *FOXP2*. What we do know is that in mice (and so, presumably, in humans) *FOXP2* is active in the brain during embryonic development.

Contrary to initial speculation, the KE family had not reverted to a 'chimp-like' version of the gene – they had a new mutation that set back their language skills. In any case, chimps, mice and most other species have a version of *FOXP2* that is remarkably similar to that of humans. But since we split from chimpanzees there have been two other mutations to the human version, each of which alters just one of the many amino acids that make up the *FOXP2* protein.

It would be fascinating to put the human version of *FOXP2* into chimps to see if it improves their powers of speech, but we cannot do that for both technical and ethical reasons. The human version has been put into mice, though. Intriguingly, the researchers observed that the genetically modified mice pups squeak slightly differently – there was a small drop in the pitch of their ultrasound squeals.

But this may be less relevant than the changes seen within the mice brains. Changes have been found in the structure and behaviour of neurons in an area called the cortico-basal ganglia circuits, for instance. Also called the brain's reward circuits, these are known to be involved in learning new mental tasks. 'If you do something and all of a sudden you get a reward, you learn that you should repeat that,' says Wolfi Enard, an evolutionary geneticist at the Max Planck Institute for Evolutionary Anthropology in Leipzig, Germany, who led the work.

Based on what we already know about these circuits, Enard thinks that in humans *FOXP2* plays a role in learning the rules of speech – that specific vocal movements generate certain sounds, perhaps, or even the rules of grammar. 'You could view it as learning the muscle sequences of speech, but also learning the sequence of "The cat the dog chased yesterday was black",' he suggests.

Enard reckons this is the best example yet found of a mutation that fuelled the evolution of the human brain. 'There's no other mutation where we have such a good idea what happened,' he says.

5) Helping hand

From the first simple stone tools, through to the control of fire and the development of writing, our progress has been dependent on our dexterity. It's not for nothing that in the science-fiction classic *2001: A Space Odyssey* Arthur C. Clarke portrayed the day an ape-man started clubbing things with an animal bone as a pivotal moment in our evolution.

Assuming alien meddling was not responsible, can our DNA shed light on our unrivalled abilities with tools? Clues come from a DNA region called HACNS1, short for human-accelerated conserved non-coding sequence 1, which has undergone 16 mutations since we split from chimps. The region is an on/off switch that seems to kick a gene into action in several places in the embryo, including developing limbs. Cutting and pasting the human version of HACNS1 into mouse embryos reveals that the mutated version is activated more strongly in the forepaw, right in the areas that correspond to the human wrist and thumb.

Some speculate that these mutations contributed to the evolution of our opposable thumbs, which are crucial for the deft movements required for tool use. In fact, chimps also have opposable thumbs, just not to the same extent as us. 'We have more fine muscle control,' Pollard says. 'We can hold a pencil, but we can't hang from the limb of a tree comfortably like a chimp.'

6) Switch to starch

Chimps and other large primates subsist mainly on fruits and leaves. These are such low-calorie foods that the animals have to forage for most of their waking hours. Modern humans get most of their energy from starchy grains or plant roots. Over the past 6 million years our diet must have undergone several shifts, when we started using stone tools, learned to cook with fire, and settled down as farmers.

Some of these changes are hard to date. There is an ongoing debate over what constitutes the first evidence for cooking hearths. And digging sticks, used to unearth tubers and bulbs, do not fossilise. An alternative way of tracking dietary changes is to look at the genes involved in digestion.

A digestive enzyme called salivary amylase plays a key role in breaking down starch into simple sugars so it can be absorbed in the gut. Humans have much higher levels of amylase in their saliva than chimpanzees: while chimps have only two copies of the salivary amylase gene (one on each of the relevant chromosome pair), humans have an average of six, with some people having as many as 15. DNA-copying errors during the production of sperm and eggs must have led to the gene being repeatedly duplicated.

To find out when the duplications happened, the gene was sequenced in people from several countries, as well as in chimps and bonobos. 'We were hoping to find a signature of selection about 2 million years ago,' says Nathaniel Dominy, a biological anthropologist at Dartmouth College in Hanover, New Hampshire, who led the amylase gene work. That is around the time our brains underwent significant growth, and one theory is that it was fuelled by a switch to a starchier diet.

But the team found the gene duplications had happened more recently – some time between 100,000 years ago and the present day. The biggest change in that period was the dawn of agriculture, so Dominy thinks the duplications happened when we started farming cereals. 'Agriculture was a signal event in human evolution,' he says. 'We think amylase contributed to it.'

It was the advent of agriculture that allowed us to live in larger settlements, which led to innovation, the cultural explosion and, ultimately, modern life. If we consider all the mutations that led to these pivotal points in our evolution, human origins begin to look like a trail of unfeasible coincidences. But that is only because we do not see the harmful mutations that were weeded out, Hawks points out. 'What we're left with is the ones that were advantageous.' It is only from today's viewpoint that the mutations that give us our current physical form appear to be the 'right' ones to have. 'It's hindsight,' says Hawks. 'When we look back at the whole process, it looks like a stunning series of accidents.'

3

Chance vs the brain

Why you can't handle the truth

Our brains are amazing. As we have seen, they can do science well enough to work out how we got here, and even appreciate the role that chance played in the process. But left to their own devices, randomness deceives them time and again. In the chapter after this one, we'll get down to the numerical nitty-gritty, the mathematics of randomness. For now, though, let's spend some time enjoying the human mind's strange relationship with chance. There's coincidence and luck to consider – where do they come from, and why do they make fools of us so often? We'll explore whether the brain can handle the demands of a casino, and whether you can learn to act randomly. And we'll look at how to make chance work for you – being lucky isn't always a matter of luck.

That's amazing – isn't it?

We love to see chance in action, attaching meaning to the kinds of correlated events that, to a statistician, have none at all. It seems to be a primal reflex, and it is likely that humans have always seen significance in insignificant things. Here, Ian Stewart and Jack Cohen take a look at the psychological catnip that is unanticipated coincidence.

The scene is Jerez Grand Prix circuit, the last race of the 1997 Formula One season. Michael Schumacher is one championship point ahead of arch-rival Jacques Villeneuve – thanks in part to brilliant tactical driving by his Ferrari teammate Eddie Irvine in the previous race. Villeneuve's Williams teammate Heinz-Harald Frentzen may well play the same game this time, so qualifying in pole position is even more critical than usual...

So what happens? Villeneuve, Schumacher and Frentzen all lap in exactly 1 minute 21.072 seconds. The astonished commentators hailed it as an amazing coincidence. Well, 'coincidence' it surely was – the lap times coincided. But was it truly amazing?

Questions like this do not just apply to sport. They turn up everywhere, trivial and important. Just how surprising was it to run into Great Aunt Lottie from Sweden in that San Francisco strip bar? Is it really unexpected that three different people at the Christmas party are wearing the same dress? And in science, how significant is a leukaemia cluster? Does a strong correlation between lung cancer and having a smoker in the family really prove that passive smoking is dangerous?

One of us, Jack Cohen, is a reproductive biologist. He was once asked to explain two very curious statistics. While in Israel, he was told that 84 per cent of the children of Israeli fighter pilots are girls. 'What is it about the life of a fighter pilot,' he was asked, 'that produces such a predominance of daughters?' The second statistic arose in connection with in-vitro fertilisation. Nowadays, IVF clinics use ultrasound to monitor ovulation, and so can determine whether an egg – and the resulting baby – comes from the left or right ovary. One clinic discovered that most of the girl babies came from the left ovary, and most of the boys from the right. A breakthrough in choosing the sex of your children? Or just a statistical freak?

It's not easy to decide. Gut feelings are worse than useless, because human intuition is poor when it comes to random events. Many people believe that lottery numbers that have so far been neglected must be more likely to come up in future. In justification they plead the 'law of averages' – everything ought to even out in the long run. But the truth is different, and not at all intuitive. Yes, in the long run, each lottery number is indeed just as likely to turn up as any other. But the lottery machine has no memory. The proportions do even out, in the long run, but you can't say in advance how long that run will be. In fact, if you choose any specific number of attempts, however large, then the best prediction is that any initial imbalance will remain unchanged.

Our intuition goes even further astray when we think about coincidences. You go to the local swimming pool, and the guy behind the counter pulls a key at random from a drawer full of keys. You arrive in the changing room and are relieved to find that very few lockers are in use... and then it turns out that three people have been given lockers next to yours, and it's all 'Sorry!' as you bang locker doors together. Or you are in Hawaii, for the only time in your life ... and you bump into the Hungarian you worked with at Harvard. Or you're on honeymoon camping in a remote part of Ireland... and you and your new wife meet your department head and his new wife, walking the other way along an otherwise deserted beach. All of which happened to Jack.

These coincidences all seem striking because we expect random events to be evenly distributed, so statistical clumps surprise us. We think that a 'typical' draw in Britain's national lottery is something like 5, 14, 27, 36, 39, 45, but that 1, 2, 3, 19, 20, 21 is far less likely. In fact, these two sets of numbers have the same probability: 1 in 13,983,816. Sequences of six random numbers are actually more likely to be clumpy than not.

How do we know this? Probability theorists tackle such questions using 'sample spaces'. A sample space contains not just the event that concerns us, but all possible alternatives. If we are rolling a die, for instance, then the sample space is 1, 2, 3, 4, 5, 6. For the UK lottery, the sample space is the set of all sequences of six different numbers between 1 and 49. A numerical value is assigned to each event in the sample space, called its 'probability', and this corresponds to how likely that event is to happen. For fair dice each value is equally likely, with a probability of 1 in 6. Ditto for the lottery, but now with a probability of 1 in 13,983,816.

Thinking about the size of the sample space is a good

way to assess how amazing an apparent coincidence really is. Take the Formula One lap times. Top drivers all usually lap at roughly the same speed, so it's reasonable to assume that the three fastest times would fall inside the same tenth-of-a-second period. At intervals of a thousandth of a second, there are 100 possible lap times for each to choose from: this list determines the sample space. Assume for simplicity that each time in that range is equally likely. Then there is a 1 in 100 chance that the second driver laps in the same time as the first, and a 1 in 100 chance that the third laps in the same time as the other two – which leads to an estimate of 1 in 10,000 as the probability of the coincidence. Low enough to be striking, but not so low that we ought to feel truly amazed. It's roughly as likely as a hole-in-one in golf.

Estimates like this help to explain astounding coincidences reported in newspapers, such as bridge players getting a 'perfect hand', each having the thirteen cards in a suit. In any one game, the chances of this happening are staggeringly small. But the number of games of bridge played every week worldwide is huge. So huge that every few weeks the actual events explore the entire possible sample space. In other words, you should expect perfect hands to turn up somewhere – as often as their small but non-zero probability predicts.

The use of sample spaces, however, is not entirely straightforward. Statisticians tend to work with the 'obvious' sample space. For that question about Israeli fighter pilots, for instance, they would naturally take the sample space to be all children of Israeli fighter pilots. But that might well be the wrong choice. Why? We often tend to underestimate the size of the sample space – and so assume that coincidences are more surprising than they really are. It is all down to a crucial factor which we call 'selective reporting' – which tends to be ignored in most conventional statistics.

That perfect hand at bridge, for instance, is far more likely to make it to the local or even national press than an imperfect one. How often do you see the headline 'Nottingham Bridge Players Get Entirely Ordinary Hand', for instance? The human brain just can't resist looking for patterns, and seizes on certain events that it considers significant, whether or not they really are. And in so doing, it ignores all the 'neighbouring' events that would help it judge how likely or unlikely the perceived coincidence actually is.

Selective reporting affects the significance of those Formula One times. If it hadn't been them, maybe the tennis scores in the US Open would have contained some unusual pattern, or the snooker breaks in that week's tournament, or the golf … Any one of those would have been reported too. But none of the failed coincidences, the ones that didn't quite happen, would have hit the headlines. If you include just ten major sporting events in the list of would-bes that weren't, that 1 in 10,000 chance comes down to only 1 in 1,000. It's like tossing a coin and turning up heads 10 times in a row.

So going back to the Israeli fighter pilots – is it just random chance or is something else happening? To answer this, conventional statistics would set up the obvious sample space (children of fighter pilots), assign probabilities to boy and girl children, and calculate the chance of getting 84 per cent girls in a purely random trial. But this analysis ignores selective reporting. Why did anyone look at the sexes of Israeli fighter pilots' children in the first place? Presumably because a clump had already caught their attention. If it had been the heights of the children of Israeli aircraft manufacturers, or the musical abilities of the wives of Israeli air traffic controllers that showed up as a clump, their clump-seeking brains would have pounced on those strange circumstances instead. The conventional statistical approach tacitly excludes many other factors that didn't clump – it ignores part of the sample space.

The human brain filters vast quantities of data, seeking things that appear unusual, and only then does it send out a conscious signal: will you look at that! The wider you cast your pattern-seeking net, the more likely it is to catch a clump. There's nothing wrong with that. But if you want to know how significant the clump is, you can't include the data that drew your attention to the clump in the first place. In a room with 20 people there will typically be one – the tallest – whose height puts them in the top few per cent in the country. But you can't then remove that person from the room, re-measure their height, and then deduce that it is surprising that you have found somebody with such an unusual height. That's what you chose them for.

This is exactly the error made in early experiments on extra-sensory perception. Thousands of subjects were asked to guess cards from a special pack of five symbols. After several weeks of selection, anyone whose success rate had been above average was invited back and tested some more. At first, these 'good guessers' seemed to have extraordinary powers. But as time went on, their success rate slowly dropped back towards the average, as if their powers were 'running down'. This happened because their initial high scores – the clumping for which they were chosen – were included in the running total. If these fortuitous scores had been excluded from the second set of tests, then the scoring rate would have dropped, immediately, to near average.

And so it is with the fighter pilots, and with the left/right ovaries. The curious figures that drew researchers' attention to these particular effects may well have been the result of selective reporting – or, much the same, selective attention. If so, then you can make a simple prediction: 'From now on, the figures will revert to fifty-fifty.' If this prediction fails, and if the results instead confirm the bias that revealed the clump, then the new data can be considered significant. But the smart money is on the prediction succeeding.

The same fallacy can arise in conventional experimental studies, say, to find out whether certain foodstuffs give you cancer. To save time, the usual way is to look at many different foods at once – fibre, fat, red meat, vegetables and so on – and see how they all compare with cancer rates. So far so good. But then you pick out the biggest entry – the food that is unusually closely correlated with cancer rates. Unless you're careful, you now forget all the other factors, and publish a paper saying that eating red meat significantly increases your chances of cancer. However, you chose the most significant from the hundred different foods you tried. Of course at least one comes out to be significant. You'd be surprised if it didn't, on statistical grounds alone, even if all the foods were chosen at random.

An alleged decline in the human sperm count may be another example of selective reporting. Niels Skakkebaek's team at Copenhagen University published the first widely accepted evidence for a decline in 1997. But it is not these researchers who are to blame for the selective reporting. In the frame for that are researchers who had contrary evidence but didn't publish it because they thought it must be wrong; journal referees who accepted papers that confirmed a decline more often than they accepted those that didn't; and the press – who strung together a whole pile of sex-related defects in various parts of the animal kingdom into a single seamless story, unaware that each individual instance had an entirely reasonable explanation that had nothing to do with falling sperm counts and often nothing to do with sex. Sexual abnormalities in fish kept in water from sewage treatment plants, for instance, are probably due to excess nitrites – which all fish-breeders know cause abnormalities of all kinds – and not to oestrogen-like compounds in the water, which would bolster the 'sperm count' story.

The message here is that when you are estimating statistical significance, you must tailor your choice of sample space to the experiment as it was actually performed, not a selectively reported version. The safest way to do this is to discard the data that led to your particular result, and repeat the experiment to get new data. But even then you must not let the coincidence, the clump, choose the sample space for you – if you do, you're ignoring the surrounding space of near-coincidences.

We decided to test this theory on a trip to Sweden. On the plane, Jack predicted that a coincidence would happen at Stockholm airport. His reason: selective reporting. If we looked hard enough, we'd be sure to find one. We got to the bus stop outside the terminal, and no coincidences had occurred. But we couldn't find the right bus, so Jack went back to the enquiries desk. As he waited, someone came up next to him – Stefano, a mathematician who normally worked in the office next door to Jack's at Warwick University, UK.

Prediction confirmed. But what we really wanted was to lay hands on a near-coincidence – one that hadn't happened, but could have been selectively reported if it had. For instance, if some other person we knew had shown up at exactly the same time, but on the wrong day, or at the wrong airport, we'd never have noticed. So for all we knew, Sweden could be overrun by acquaintances of ours, making it virtually impossible for us not to bump into one of them at almost any moment. Near-coincidences, by definition, are hard to observe. But we happened to mention all of the above to Ian's friend Ted, who visited soon after we got back to England. 'Stockholm?' said Ted: 'When?' And we told him. 'Which hotel?' 'The Birger Jarl.' 'Funny, I was staying at the Birger Jarl one day later than you!' Had we travelled one day later, then, we wouldn't have met Stefano – but we would have met Ted. Selective reporting would have ensured that we only told our friends about the one that actually happened.

Probability theory assesses how likely an event is in comparison with others that could have happened, but somehow or another didn't. Our intuition for probabilities is poor because the feature-detection system in our brains notices only the things that happen. In the world around us, every event is unique. Every meeting, every sex ratio, every bridge hand. 'What, your telephone number is nearly the same as your car registration? How amazing!' But when you realise that a typical citizen has several dozen significant numbers (address, postcode, PINs, mobile, credit cards …) a chance resemblance between two of them is surely mundane, rather than remarkable.

What we must not do, then, is to look back at past events and find significance in the inevitable few that look odd. That is the way of the pyramidologists and the tea-leaf readers. Every pattern of raindrops on the pavement is unique. We're not saying that if one such pattern happens to spell your name this is not to be wondered at – but if your name had been written on the pavement in Beijing during the Ming dynasty, at midnight, nobody would have noticed. It's no use looking at past history when assessing significance: you need to look at all the other things that might have happened instead.

Every actual event is unique. Until you place that event in a category, you can't work out which background to view it against. Until you choose a background, you can't estimate the event's probability. On the other hand, if something that seems spooky to you really does turn out to have a small sample space, that's when you should be really amazed.

The luck factor

Do you feel lucky? It's easy to see luck as something the universe forces on you. But Richard Wiseman has carried out an intriguing series of experiments that show you are not fortune's fool, as Shakespeare's Romeo would have it. Instead, as he makes clear here (Wiseman, not Shakespeare – or Romeo), luck is about whether you are poised to exploit random events for personal gain. It turns out, you really do make your own luck.

People think about luck in one of two conflicting ways. Either we all make our own luck, or it is an uncontrollable, overbearing random force. But which is it? To answer this, I set out to investigate the science of luck, and to discover whether it is possible to help people make more of their chances. The results give a fascinating and surprising insight into the secret science behind our everyday lives.

Academic interest in luck stretches back to the early 1930s, when psychologist Joseph Banks Rhine at Duke University in North Carolina carried out some of the first experiments into the possible existence of extra-sensory perception. One day, a professional gambler visited Rhine and claimed that he could control his luck by influencing the roll of dice in casinos. Intrigued, Rhine and his colleagues staged a series of experiments in which they asked people to rate the degree to which they considered themselves lucky or unlucky, and then to complete standard tests of psychic ability, such as trying to determine the order of a shuffled deck of cards. The results proved inconclusive and researchers eventually lost interest in the science of luck.

In the early 1990s I came across this initial work, and decided to conduct my own investigation. I placed advertisements in newspapers and magazines, asking people who considered themselves exceptionally lucky or unlucky to contact me. Over the next few years we received replies from around a thousand people in the UK. Their experiences shared remarkable similarities. Lucky people were always in the right place at the right time, had more than their fair share of opportunities, and lived a 'charmed life'. Unlucky people encountered little but failure and misery, and never seemed to get a break. Fascinated by their stories, I set out to discover why they experienced such different lives.

The first stage was much like the early work conducted by parapsychologists. We asked participants to try to predict the outcome of genuinely random events, such as the National Lottery. Time and again there was no difference between the performance of the lucky and unlucky groups, with the scores of both being consistent with chance.

So we turned our focus onto the messy complexities of everyday life. We wondered whether many of the seemingly chance events that happen in our lives are actually the result of the way we think and behave. Perhaps lucky and unlucky people are unconsciously creating their own good and bad fortune. For example, perhaps lucky people are especially outgoing and have more 'chance' encounters simply because they meet more people. Or maybe unlucky people are especially pessimistic and so less likely to succeed because they don't persevere in the face of failure.

To explore these ideas, we carried out a series of studies. In one small demonstration, for example, we placed a five-pound note on the street outside a coffee shop where we'd invited a self-professed lucky and unlucky person to attend. The lucky man saw the money, took it into the coffee shop and within minutes had started a conversation with a successful businessman (who was really a stooge of ours) and bought him a coffee. By contrast, our unlucky woman missed the five-pound note, bought a coffee for herself and drank it alone. When asked later how their mornings had gone, the woman described an uneventful time while the man was upbeat about his enjoyable time.

From a series of larger experiments and questionnaires of other people who rated themselves on a 'luck scale', big differences emerged between the lucky and unlucky. We found that lucky people are skilled at creating and noticing opportunities by, for example, networking, adopting a relaxed attitude to life and being open to new experiences. Also, when faced with a complex and important issue in their lives, lucky people tend to make more effective decisions by listening to their intuition and gut feelings. Lucky people are certain that the future is going to be full of good fortune. These expectations become self-fulfilling prophecies because they help lucky people to motivate themselves and those around them. Finally, lucky people are especially resilient, and when ill-fortune strikes they are able to cope by, say, imagining how things could have been worse or taking control of their situation.

In the final phase of research we examined whether it was possible to improve people's good fortune by getting them to think and behave like a lucky person. This type of change is a very thorny issue in psychology. Some researchers believe that the fundamental aspects of personality are hard-wired into our brains, and tend not to change. Others, including me, believe that in some aspects of personality real change is possible.

During our studies, we assembled a group of people who did not consider themselves lucky or unlucky, and asked them to carry out a series of simple exercises designed to encourage them to think like a lucky person. For example, the group was asked to spend a few minutes every day focusing on the positive aspects of their lives, to connect more with others and to adopt a more relaxed attitude to life. After a few months we assessed their lives, including happiness, physical health and, of course, how lucky they thought they were. Overall, participants had become happier, healthier and luckier. In short, changing the way you think and behave can create real and lasting improvements to your life.

I am regularly asked to speak about my research. From high-tech companies to elite sportspeople, a surprisingly wide variety of organisations and individuals are now open to the new science of luck. In the UK, where I live, many seem drawn to this idea – perhaps because Britons rate modesty highly, while 'behaving lucky' means being outgoing and making the most of your contacts and chances. (In the US, people seem somewhat less inhibited about this.) Whatever the reason, my hope is that we will all soon come to see luck as it really is – a skill that can be learned and mastered.

Throughout history, people have realised how a few seconds of ill-fortune can cause years of hardship: most people struck by lightning, for example, suffer some form of permanent disability. Good luck, on the other hand, can save an enormous amount of hard work. Take actress Charlize Theron, whose career took off after years of struggle when she threw a hissy fit in a bank. Her histrionics impressed a showbiz agent who happened to be in line behind her.

Because of stories like these, many have tried to increase their luck by carrying out superstitious rituals and carrying lucky charms. This approach has been unsuccessful because it is based on an outdated, inaccurate way of thinking about the issue. The time has come to adopt a more rational, and scientific, approach to good fortune: get out there and make your own luck.

Go crazy!

If your brain is not good at turning random events into opportunities, perhaps you can learn to create opportunities from randomness? Switch off your inner organiser, let unpredictability rule, and you could become the unbeatable champion of the world's favourite hand game: rock-paper-scissors. This is not advice from a random stranger, it's from a certain Michael Brooks.

You know the score: paper wraps rock, rock blunts scissors, scissors cut paper … It's just a trivial way of making decisions about whose round it is at the bar, who gets the TV remote, that kind of thing. It's something like tossing a coin, right?

You couldn't be more wrong. Rock, paper, scissors (RPS) – also known as RoShamBo – is a startling game of strategy that reveals both the fickleness and the limitations of the human mind. There are RPS world championships worth big money, fiercely contested tournaments to find the best RPS computer programs, and heated arguments over which is the optimal RPS strategy. When millions of dollars have been made on the throw of a hand, it is hard to argue this is an insignificant debate. So, how do you win at RPS?

From a mathematical perspective, RPS is a function known as an intransitive relation, which means it creates a loop of preferences that has no beginning and no end, defying standard notions of hierarchy. Though each item is better than some other item, it is impossible to define what is 'best', and this makes it interesting to mathematicians. 'It makes you think precisely about what you mean by "is better than",' says John Haigh, a mathematician at the University of Sussex in the UK. 'Context is everything.'

Given the interest among mathematicians, it was almost inevitable that computer programmers would get involved and try to produce the ultimate player. According to game theory, the optimal strategy is straightforward: make your throws random. If no one can guess what you're going to play, they can't devise a winning strategy against you. That may sound like a simple thing to do, but it isn't – not even for computers – as David Bolton, an author and app developer, has demonstrated.

For a while, Bolton, an RPS enthusiast, ran a computer RPS league. The competitors supplying their game-playing code came from as far afield as the Philippines,

South Africa, Sweden and China, and their programs, or bots, used a wide range of strategies. Surprisingly, the least successful bots were those that seemed to make their choice based on nothing more than random numbers. 'These all tended to be at the bottom of the league,' Bolton says.

The explanation must be that these poor performers were not truly random. If there are any patterns at all, well-programmed bots will pick them out – and work out how to exploit them. They might, for instance, analyse their opponents' previous moves to find a pattern and thus work out the most likely next move.

Though competitions between programs are a challenge for the programmers, they are of limited interest to everyone else, says Perry Friedman, who created RoShamBot, one of the first RPS bots. Computer RPS players are simply too good. 'It's much more interesting to find games that play well against people,' Friedman says. So when Friedman created RoShamBot, he deliberately refrained from making it invincible. While the program is powerful, its charm, he says, is that it doesn't just mash you into a pulp (you can play against RoSham-Bot via http://ro-sham-bot.appspot.com/).

Since graduating from Stanford University, Friedman has worked as a programmer for IBM and Oracle and as a professional poker player. In the latter pursuit, playing RPS against other humans has been a big help, Friedman says, because live-action RPS teaches you about the peculiarities of human thought. In RPS, the golden rule is to be unpredictable, but without extensive training humans are hopeless at this. 'People tend to fall into patterns,' Friedman says. 'They tell themselves things like, "I just went rock twice, so I shouldn't do rock a third time, because that's not random".'

Worse, people tend to project patterns on their opponents. 'They see patterns where there are none,' Friedman says. This, he adds, is a major source of complaints in online gaming: when players lose because of something they perceive as a too-lucky dice throw, say, they think the computer they are playing against must be rigged. 'What are the odds double-six came up right when he needed it?' players ask. The thing is, as Friedman points out, 'They don't notice all the times it didn't come up.'

If you are going to win at RPS, Friedman's advice is to think – but not too much. Of course you want to randomise your throws, but once the game is under way you should look for patterns. If your opponent is human, the chances are he or she works – consciously or unconsciously – with a sequence in their head. Spot it, and they are toast.

Another tip is don't throw rock in your first game. This strategy won the auction house Christie's millions of dollars in 2005 when a wealthy Japanese art collector couldn't decide which firm of auctioneers should sell his corporation's collection of Impressionist paintings. He suggested they play RPS for it. Christie's asked for suggestions from their employees, one of whom turned out to have daughters who played RPS in the schoolyard. 'Everybody expects you to choose rock,' the girls said, so their advice was: go for scissors. Christie's acted on this expert tip, while rival auction house Sotheby's went for paper – and lost the business.

Scissors is still a good starting throw even if you are

playing against someone experienced: they won't go for rock because that's seen as a rookie move, so the worst you are likely to do is tie. Once things are under way, different techniques come in. You could try the double bluff, where you tell your opponent what you're going to throw – then do it: no one believes you'll do it, so they won't play the throw that beats your throw. Then, if your mind goes blank, play the throw that would have been beaten by your opponent's previous throw: some kind of subconscious activity seems to encourage players – especially those who are not feeling at the top of their game – to aim to beat their own preceding throw.

When all else fails, the rule is 'go with paper', because rock comes up more often than it would by chance. In 1998, Mitsui Yoshizawa, a mathematician at Tokyo University of Science, studied throws from 725 people and found that they threw rock 35 per cent of the time. Paper came in at 33 per cent and scissors at 31 per cent. For a while, Facebook had an online game called Roshambull which logged 10 million throws in over 1.6 million games. Here the statistics were 36 per cent rock, 30 per cent paper and 34 per cent scissors. 'Players clearly have a slight preference for rock, and that affects the distribution of all the plays,' says Graham Walker of the World RPS Society, whose website offers online training in the game. This pleases him, since it shows how winning something like the world RPS championship involves skill, not luck. 'Given people's preference for rock, it is impossible to claim that RPS is a game of chance,' he says.

So there you go: now you know how to get your own way more often than not. Do a little study, practise against an online trainer, then, wide-eyed, make what looks like an innocent suggestion: shall we settle this with rock, paper, scissors?

The chips are down

Nowhere is your pattern-hunting brain more dangerous than when let loose in a house of random chance. At least, that's what an intrepid New Scientist *reporter found when she decided to apply her mathematical skills in a local casino. But it's not just in gambling that maths can help your brain deal with the odds. It could even help you choose someone to marry. Here's Helen Thomson's experience.*

In 2004, Londoner Ashley Revell sold his house, all his possessions and cashed in his life savings. It raised £76,840. He flew to Las Vegas, headed to the roulette table and put it all on red.

The croupier spun the wheel. The crowd held its breath as the ball slowed, bounced four or five times, and finally settled on number seven. Red seven.

Revell's bet was a straight gamble: double or nothing. But when Edward Thorp, a mathematics student at the Massachusetts Institute of Technology, went to the same casino some 40 years previously, he knew pretty well where the ball was going to land. He walked away with a profit, took it to the racecourse, the basketball court and the stock market, and became a multimillionaire. He wasn't on a lucky streak. He was using his knowledge of mathematics to understand, and beat, the odds.

No one can predict the future, but the powers of probability can help. Armed with this knowledge, a high-school mathematics education and £50, I headed off to find out how Thorp, and others like him, have used mathematics to beat the system. Just how much money could probability make me?

When Thorp stood at the roulette wheel in the summer of 1961 there was no need for nerves – he was armed with the first 'wearable' computer, one that could predict the outcome of the spin. Once the ball was in play, Thorp fed the computer information about the speed and position of the ball and the wheel using a micro-switch inside his shoe. 'It would make a forecast about a probable result, and I'd bet on neighbouring numbers,' he told me.

Thorp's device would now be illegal in a casino, and in any case getting a computer to do the work wasn't exactly what I had in mind. Could I still beat the house by spinning the wheel? Possibly, but only with deep pockets and a strong faith in probability theory.

Each spin of a roulette wheel is independent. There are various bets I could make, such as betting on individual numbers, or on a single colour: red or black. I could even bet on a 'split dozen', with a chip on the junction of two columns. As a novice, I want to keep things simple, though, and with this in mind my best option is probably to bet on a single colour. However, the odds of winning on a single spin are less than 50:50.

That's because of a small twist: the number zero, which is green. This skews the odds away from 50:50 black or red towards odds that mean the house will always win in the long run – I will win only 48.6 per cent of the time. And that's in Europe; American roulette wheels have two green zeroes so that the house wins even faster. In America, I'd win 47.4 per cent of the time on average.

If I want to maximise my chances of making a profit, I could use the strategy of repeatedly betting on a colour, while doubling my bet on the next spin if I should lose. But I'd need a big pot of cash to stay in the game: a losing streak will escalate my bets very quickly. Seven unlucky spins on a £10 starting bet will have me parting with a hefty £1,280 on the next. What's more, casinos operate a maximum bet policy, so even if I had deep pockets, I couldn't play for ever. And even if I were to have a lucky streak, my winnings wouldn't escalate in the same way as my losses: any win only makes a profit equal to the original stake, and every loss costs me ever more. So, although this strategy is logically sound, it is hugely risky, with only a very small gain. The roulette wheel is likely to keep on taking my money far longer than I can remain solvent.

With that in mind, I turned my back on roulette and followed Thorp into the card game blackjack. In 1962 he published a book called *Beat the Dealer*, which proved what many had long suspected: by keeping track of the cards, you can tip the odds in your favour. He earned thousands of dollars putting his proof into practice.

The method is now known as card counting. So does it still work? Could I learn to do it? And is it legal?

'It's certainly not illegal,' Thorp assures me. 'The casino can't see inside your head – yet.' What's more, after a brief tutorial, it doesn't sound too difficult. 'If you went into any casino that had basic blackjack rules, learned the method of card counting that I've taught you, you'd have a modest advantage without much effort,' says Thorp.

Basic card counting is simple. Blackjack starts with each player being dealt two cards face up. Face cards count as 10 and the ace as 1 or 11 at the player's discretion. The aim is to have as high a total as possible without 'busting' – going over 21. To win, you must achieve a score higher than the dealer's. Cards are dealt from a 'shoe' – a box of cards made up of three to six decks. Players can stick with the two cards they are dealt or 'hit' and receive an extra card to try to get closer to 21. If the dealer's total is 16 or less, the dealer must hit. At the end of each round used cards are discarded.

The basic idea of card counting is to keep track of those discarded cards to know what's left in the shoe. That's because a shoe rich in high cards will slightly favour you, while a shoe rich in low cards is slightly better for the dealer. With lots of high cards still to be dealt you are more likely to score 20 or 21 with your first two cards, and the dealer is more likely to bust if his initial cards are less than 17. An abundance of low cards benefits the dealer for similar reasons.

If you keep track of which cards have been dealt, you can gauge when the game is swinging in your favour. The simplest way is to start at zero and add or subtract according to the dealt cards. Add 1 when low cards (two to six) appear, subtract 1 when high cards (10 or above) appear, and stay put on seven, eight and nine. Then place your bets accordingly – bet small when your running total is low, and when your total is high, bet big. This method can earn you a positive return of up to 5 per cent on your investment, says Thorp.

After a bit of practice at home, I head off to my nearest casino. Trying to blend in among the rich young things, the shady mafia types and the glamorous cocktail waitresses was one thing; counting cards while trying to remain calm was another. 'If they suspect that you're counting cards, they'll ask you to move to a different game or throw you out completely,' one of the casino's regulars tells me.

After a few hours I begin to get the hang of it, and eventually walk away with a profit of £12.50 on a total stake of £30. The theory is good, but in practice it's a lot of effort for a small return. It would be a lot easier if I could just win the lottery. How can I improve my chances there? There's a lesson to be learned here from the experience of one Alex White.

The evening of 14 January 1995 was one that White will never forget. He matched all six numbers on the UK National Lottery, with an estimated jackpot of a massive £16 million. Unfortunately, White (not his real name) only won £122,510 because 132 other people also matched all six numbers and took a share of the jackpot.

There are dozens of books that claim to improve your odds of winning the lottery. None of them works. Every combination of numbers has the same odds of winning as any other – 1 in 13,983,816 in the case of the UK 'Lotto' game. But, as White's story shows, the fact that you might have to share the jackpot suggests a way to maximise any winnings. Your chances of success may be tiny, but if you win with numbers nobody else has chosen, you win big.

So how do you choose a combination unique to you? You won't find the answer at the National Lottery head-quarters – they don't give out any information about the numbers people choose. That didn't stop Simon Cox, a mathematician at the University of Southampton, UK, from trying. In 1998, Cox worked out UK lottery players' favourite figures by analysing data from 113 lottery draws. He compared the winning numbers with how many people had matched four, five or six of them, and thereby inferred which numbers were most popular.

And what were the magic numbers? Seven was the favourite, chosen 25 per cent more often than the least popular number, 46. Numbers 14 and 18 were also popular, while 44 and 45 were among the least favourite. The most noticeable preference was for numbers up to 31. 'They call this the birthday effect,' says Cox. 'A lot of people use their date of birth.'

Several other patterns emerged. The most popular numbers are clustered around the centre of the form people fill in to make their selection, suggesting that players are influenced by its layout. Similarly, thousands of players appear to just draw a diagonal line through a group of numbers on the form. There is also a clear dislike of consecutive numbers. 'People refrain from choosing numbers next to each other, even though getting 1, 2, 3, 4, 5, 6 is as likely as any other combination,' says Cox. Numerous studies on the US, Swiss and Canadian lotteries have produced similar findings.

To test the idea that picking unpopular numbers can maximise your winnings, Cox simulated a virtual syndicate that bought 75,000 tickets each week, choosing its numbers at random. Using the real results of the first 224 UK lottery draws, he calculated that his syndicate would have won a total of £7.5 million – on an outlay of £16.8

million. If his syndicate had stuck to unpopular numbers, however, it would have more than doubled its winnings.

So the strategy is clear: go for numbers above 31, and pick ones that are clumped together or situated around the edges of the form. Then if you match all six numbers, you won't have to share with dozens of others.

Unfortunately, probability also predicts that you won't match all six numbers until the 28th century. I bought a ticket and chose some of Cox's least popular numbers: 26, 34, 44, 46, 47 and 49. Not one of them came up. So I headed for the bookmaker.

Although it would be nearly impossible to beat a seasoned bookie at his own game, play two or three bookies against each other and you can come up a winner. So claims John Barrow, professor of mathematics at the University of Cambridge, in his book *100 Things You Never Knew You Never Knew*. Barrow explains how to hedge your cash around different bookies to ensure that whatever the outcome of the race, you make a profit.

Although each bookie will stack their own odds in their favour, thus ensuring that no punter can place bets on all the runners in a race and guarantee a profit, that doesn't mean their odds will necessarily agree with those of a different bookie, says Barrow. And this is where gamblers can seize their chance.

Let's say, for example, you want to bet on one of the highlights of the British sporting calendar, the annual university boat race between old rivals Oxford and Cambridge. One bookie is offering 3 to 1 on Cambridge to win and 1 to 4 on Oxford. But a second bookie disagrees and has Cambridge evens (1 to 1) and Oxford at 1 to 2.

Each bookie has looked after his own back, ensuring that it is impossible for you to bet on both Oxford and Cambridge with him and make a profit regardless of the result. However, if you spread your bets between the two bookies, it is possible to guarantee success (see diagram on page 85 for details). Having done the calculations, you place £37.50 on Cambridge with bookie 1 and £100 on Oxford with bookie 2. Whatever the result, you make a profit of £12.50.

Simple enough in theory, but is it a realistic situation? Yes, says Barrow. 'It's very possible. Bookies don't always agree with each other.'

Guaranteeing a win this way is known as 'arbitrage', but opportunities to do it are rare and fleeting. 'You are more likely to be able to place this kind of bet when there are the fewest possible runners in a race, therefore it is easier to do it at the dogs, where there are six in each race, than at the horses, where there are many more,' says Barrow.

Even so, the mathematics is relatively simple, so I decided to try it out online. The beauty of online betting is that you can easily find a range of bookies all offering slightly different odds on the same race. 'There are certainly opportunities on a daily basis,' says Tony Calvin of online bookie Betfair. 'It's not necessarily risk-free because you might not be able to get the bet you want exactly when you need it, but there are certainly people who make a living out of arbitrage.'

After persuading a few friends to help me try an online bet, we followed a race, each keeping track of a horse and the odds offered by various online bookies. Keeping track of the odds to spot arbitrage opportunities was hard enough. Working out what to bet and when was, unsurprisingly, even harder. Arbitrage is not for the uninitiated.

However, it's still quite addictive, especially when you get tantalisingly close to finding a winning combination. And that's the problem with gambling – even when you have got mathematics on your side, it's all too easy to lose sight of what you could lose. Fortunately, that's the final thing that probability can help you with: knowing when to stop.

Everything in life is a bit of a gamble. You could spend months turning down job offers because the next one might be better, or keep laying bets on the roulette table just in case you win. Knowing when to stop can be as much of an asset as knowing how to win. Once again, mathematics can help.

If you have trouble knowing when to quit, try getting your head around 'diminishing returns' – the optimal stopping tool. The best way to demonstrate diminishing returns is the so-called marriage problem. Suppose you are told you must marry, and that you must choose your spouse out of 100 applicants. You may interview each applicant once. After each interview you must decide whether to marry that person. If you decline, you lose the opportunity for ever. If you work your way through 99 applicants without choosing one, you must marry the 100th. You may think you have 1 in 100 chance of marrying your ideal partner, but the truth is that you can do a lot better than that.

If you interview half the potential partners then stop at the next best one – that is, the first one better than the

How to beat the bookies
Suppose there is a race with N runners
You can always make a profit if Q is less than 1,

where $Q = \dfrac{1}{(a_1+1)} + \dfrac{1}{(a_2+1)} + \dots + \dfrac{1}{(a_n+1)}$; a_1 is the odds on runner 1, a_2 is the odds on runner 2, etc

If Q <1 there is an arbitrage opportunity. You can take advantage of it by gambling $\left(\dfrac{\frac{1}{a_1+1}}{Q}\right)$ of your money on runner 1, $\left(\dfrac{\frac{1}{a_2+1}}{Q}\right)$ on runner 2, and so on

As a simple example, take the Oxford and Cambridge University boat race

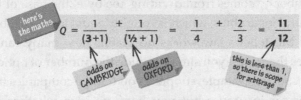

Bookie 1 is offering 3 to 1 on Cambridge to win — and 1 to 4 on Oxford to win

Bookie 2 has Oxford at 1 to 2 — and Cambridge evens (1 to 1)

You can guarantee profit

Bet on Cambridge with bookie 1 and Oxford with bookie 2

here's the maths

$$Q = \frac{1}{(3+1)} + \frac{1}{(\frac{1}{2}+1)} = \frac{1}{4} + \frac{2}{3} = \frac{11}{12}$$

odds on CAMBRIDGE

odds on OXFORD

this is less than 1, so there is scope for arbitrage

But how much to bet?

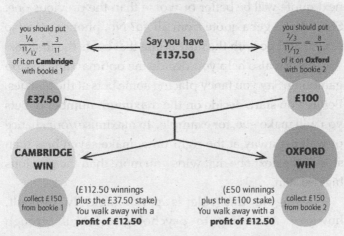

you should put $\dfrac{1/4}{11/12} = \dfrac{3}{11}$ of it on **Cambridge** with bookie 1

Say you have **£137.50**

you should put $\dfrac{2/3}{11/12} = \dfrac{8}{11}$ of it on **Oxford** with bookie 2

£37.50

£100

CAMBRIDGE WIN

collect £150 from bookie 1

(£112.50 winnings plus the £37.50 stake) You walk away with a **profit of £12.50**

OXFORD WIN

collect £150 from bookie 2

(£50 winnings plus the £100 stake) You walk away with a **profit of £12.50** best person you've already interviewed – you will marry the very best candidate about 25 per cent of the time. Once again, probability explains why. A quarter of the time, the second-best partner will be in the first 50 people and the very best in the second. So 25 per cent of the time, the rule 'stop at the next best one' will see you marrying the best candidate. Much of the rest of the time, you will end up marrying the 100th person, who has a 1 in 100 chance of being the worst, but hey, this is probability, not certainty.

You can do even better than 25 per cent, however. John Gilbert and Frederick Mosteller of Harvard University proved that you could raise your odds to 37 per cent by interviewing 37 people then stopping at the next best. The number 37 comes from dividing 100 by e, the base of the natural logarithms, which is roughly equal to 2.72. Gilbert and Mosteller's law works no matter how many candidates there are – you simply divide the number of options by e. So, for example, suppose you find 50 companies that offer car insurance but you have no idea whether the next quote will be better or worse than the previous one. Should you get a quote from all 50? No, phone up 18 (50 ÷ 2.72) and go with the next quote that beats the first 18.

This can also help you decide the optimal time to stop gambling. Say you fancy placing some bets at the bookies. Before you start, decide on the maximum number of bets you will make – 20, for example. To maximise your chance of walking away at the right time, make seven bets then stop at the next one that wins you more than the previous biggest win.

Sticking to this rule is psychologically difficult, however. According to psychologist JoNell Strough at West Virginia University in Morgantown, the more you invest, the more likely it is that you will make an unwise decision further down the line.

This is called the sunk-cost fallacy, and it reflects our tendency to keep investing resources in a situation once we have started, even if it's going down the pan. It's why you are more likely to waste time watching a bad movie if you paid to see it.

So if you must have a gamble, use a little mathematics to give you a head start, or at least to tell you when to throw in the towel. Personally I think I'll retire. Overall I'm £11.50 up – a small win at the casino offset by losing £1 on my lottery ticket. It was a lot of effort for little more than pocket change. Maybe I should have just put it all on red.

The prepared mind

Scientists can study luck, but they can also use a bit of it from time to time. We've all heard those stories about serendipitous discoveries at the laboratory bench; what's not clear is just how often they happen – and whether they really are as lucky as they sound. As it turns out, the scientists involved often don't get the credit they deserve. Bob Holmes tells us why.

If you want to make petunias a deeper purple, you could just add an extra pigment gene, right? Wrong: the extra gene turns the flowers white. This surprising finding was discovered independently in the early 1990s by two plant biologists, Richard Jorgensen in the US and Joseph Mol in the Netherlands. Neither dismissed the finding as an error. They had an inkling they'd found something big, and they had: an entirely new way in which cells regulate gene expression, now called RNA interference. RNAi has since been the subject of a Nobel Prize, has been used to save lives and promises to save many more.

The discovery is by no means the only example of good luck in science. Percy Spencer, an engineer at the US company Raytheon, was working on a radar set in 1945 when he noticed that a candy bar in his pocket was melting. That observation led, two years later, to Raytheon introducing the first commercial microwave oven. In 1976, chemist Shashikant Phadnis's boss asked him to test a chlorinated sugar compound being studied as a potential insecticide. Phadnis misheard it as a request to 'taste' the stuff – a potentially scary mistake to make in his line of work – and found it extremely sweet. We now know it as the sweetener Sucralose. Viagra was a failing drug for heart conditions before someone noticed an interesting and highly marketable side effect.

Examples like these show that chance plays a role, often a dramatic one, in the progress of science. Yet how much do we really know about its contribution? Its influence would be easier to gauge if luck in science could be defined more precisely: is it like buying a winning lottery ticket – something anyone can do – or was Louis Pasteur right to say that 'chance favours only the prepared mind'? At least one academic thinks not only that Pasteur was correct, but that it is now possible to train minds to be receptive to the subtle signs of chance.

On the question of how big a role chance plays in

science, opinions differ widely. 'There are not so many stories about serendipity. Basically, you have a couple of dozen, but in the scientific literature over the last 200 years there are so many discoveries from just plain hard work,' says Jacob Goldenberg, an innovation researcher at the Arison School of Business in Herzliya, Israel. 'If you tried to assess the ratio between serendipity-based discovery and not, I would say less than half a per cent were the result of serendipity. But we like these stories.'

Others think chance's role is more significant. 'As a social scientist, every decent idea I've ever had I had no idea about until I started doing the research and it didn't turn out the way I expected,' says Harry Collins, a sociologist of science at Cardiff University, UK. If we under-estimate the good luck factor, it's in part to do with the scale of its impact. 'I would think little surprises are there often, and big surprises are rarer,' says Michael Gorman, a social psychologist of science at the University of Virginia in Charlottesville.

One reason for the divergent views is the difficulty in defining what counts as chance. All of life, after all, is a walk down branching paths, and the choice at each fork often hangs on chance events: having an inspiring science teacher in school, an office mate who happens to know a useful tidbit of information, an experiment that improbably works out well. All of this involves chance – but it doesn't necessarily mean that discoveries happen by chance.

One of the hottest areas of neurobiology, for example, is optogenetics, which allows researchers to control the behaviour of groups of neurons with great precision.

While working at Stanford University in California, Ed Boyden and his colleagues discovered a key technique in the field, the use of light-sensitive proteins from algae to trigger electrical activity in neurons. He and his co-workers (already like-minded – the first stroke of luck) had been thinking for years about using light to control neurons. Then they stumbled across the algal studies (more good luck) and decided to try inserting the algal genes into mouse cells.

'It kinda worked on the first try,' recalls Boyden, now at the Massachusetts Institute of Technology Media Lab. 'Who would have known that these molecules from algae, which are very different organisms, would work in neurons? That was also serendipitous.' As they later learned, they were even luckier than they knew: the algal protein requires another molecule to work properly, and mammalian brains just happen to produce it for an unrelated reason.

Even so, serendipity was only half the story here. The idea of controlling neurons was one Boyden and his colleagues were keen on; in Pasteur's parlance, their minds were 'prepared'.

Perhaps the most iconic example of chance in science is Alexander Fleming's discovery of penicillin. In 1928, a stray fungal spore landed in a discarded bacterial culture in his lab at St Mary's Hospital, London. When Fleming looked at it weeks later, he noticed a ring around the growing fungal colony where something had killed the bacteria nearby. That something was eventually identified as penicillin.

Yet Fleming's finding did not pop out of a vacuum.

Other scientists over the preceding century, including Pasteur, had noticed that moulds inhibit bacterial growth. Fleming himself had spent years looking for bacteria-killing compounds and had already found one – lysozyme, an enzyme he isolated from the snot of a person with a cold. Fleming's prepared mind connected the dots, but even so it was another decade before other researchers, Howard Florey and Ernst Chain, figured out how to turn the mould into a drug.

Discoveries like these are often called 'pseudo-serendipity' – the scientists knew what they were looking for but found the answer in an unexpected place. The writer Arthur Koestler vividly described such finds as 'arrivals at the right destination by the wrong boat'. Taken to extremes, this approach can pretty much remove the element of chance from discovery. The inventor Thomas Edison, for example, tested hundreds of materials before he found the right filament for his electric light bulb, and pharmaceutical companies now systematically screen hundreds of thousands of substances looking for new drugs. When such an 'Edisonian materials dragnet', as Gorman puts it, turns up something useful, that's a testament to hard work more than luck, he says.

In contrast, true serendipity happens when researchers stumble across something entirely unexpected, as in the discovery of microwave heating or Sucralose. Here luck plays a much more obvious role – although every case still needs an alert observer to notice the anomaly, not discount it as an error, and turn it into a useful result.

Some examples, though, fall in between. Take the case of the scientist at the chemicals giant 3M who was trying to create a super-strong adhesive but ended up with a super-weak one. Years later, a colleague decided it was just the thing to stop place-markers falling out of his hymn book in church. That inspiration spawned Post-it Notes.

This sort of accident turns out to be fairly common in the annals of innovation. When Goldenberg studied the origin of 200 important inventions, he found that in about half the cases, the old saying had it backwards: invention was actually the mother of necessity. 'First they found the invention, then they discovered the need,' he says. That makes the final product not exactly an accidental discovery. It's more a matter of finding the best way to play the hand you've been dealt.

'It's much easier to find a function for an existing form than the other way around,' says Goldenberg. 'People are very creative when you have a form.' He points to the example of Vaseline, which has its roots in a dark sludge left over from oil processing. Only when chemists began looking for an application did they discover they could use the purified jelly to help burns heal.

Luck clearly helps some technologies bloom, but its impact on the broader world of scientific discovery is not so clear. Certainly, there are anecdotes, but the same few turn up time and again. No one seems to have made a systematic survey of scientific discoveries to measure how often chance plays a large part.

Indeed, such a survey may be almost impossible to do properly, says Dean Keith Simonton, a psychologist at the University of California at Davis who studies creativity. Scientific papers generally don't mention what inspired their findings, making it difficult to reconstruct the role of chance. Besides, chance may be inextricably intertwined with hard work, making it difficult to weigh the relative contribution of each. 'Even if we accept Newton's falling apple experience as valid, how much of the *Principia* should be credited to serendipity?' he asks.

Perhaps the most direct attempt to quantify scientific serendipity came two decades ago, when Juan Miguel Campanario of the University of Alcalá, Spain, surveyed 205 of the most highly cited scientific papers of the 20th century and found that 17 of them, or 8.3 per cent, mentioned some kind of serendipity contributing to the findings. This probably underestimates the true frequency, however, since not every author is likely to mention their good fortune in print.

Even if there's little certainty about how common serendipity is in science, there is broad agreement that more of it is a good thing – if only because it leads to more original discoveries. 'If you're working on something where all you have to do is be smart and work hard, chances are somebody's already found it,' says Boyden. 'So we're often trying to do things to deliberately encourage serendipity.'

Boyden has made something of a cottage industry out of wooing Lady Luck. He teaches a course at MIT on nurturing serendipity, in which he asks each group of students to systematically set out to revolutionise one area of science. 'I think we've learned enough now about how to orchestrate serendipity that maybe we should teach it,' he says.

Boyden's first rule for making your own luck in research: list all possible ideas. That's not as silly as it sounds, he argues. The trick is to subdivide the universe of possibilities into either/or options, and do it over and over again. If you're looking for a novel way to image the brain optically, for example, you could either detect photons within the brain or wait for them to leave the brain and detect them outside. If you're doing it within the brain, you could use either active electronics or a passive detector. And so on. He calls this approach a 'tiling tree' because it branches like a tree and covers the entire 'idea space' like tiles on a floor.

In effect, it's an Edisonian dragnet for ideas. 'You can subdivide into smaller and smaller categories, but you never lose any possible ideas. At the very ends of these branches are things you could try out,' says Boyden. That step is where serendipity might appear.

Boyden's second tip is to range widely. His own research group includes engineers, physicists, neuroscientists, chemists, mathematicians, and more. This diversity increases the odds that someone will make an unexpected conceptual connection. In the same vein, it's good to work on more than one thing at once, which also boosts the likelihood of cross-pollination. This was a key source of Thomas Edison's creativity, for example. In a study of the chronology of all 1,093 of Edison's patents, Simonton found that the more subjects he was working on, the higher his output of patents.

A more controversial way to encourage serendipity – especially the big discoveries that open up whole new fields of science – is simply to find the smartest, most creative thinkers and give them unrestricted funding to get on with it.

That's what used to happen at legendary research centres like Bell Labs, and still happens to some extent at Google, for example, which allows its engineers to spend 20 per cent of their time on side projects. Back in the 1980s, oil giant BP also funded a blue-skies research initiative whose goal was to seek out the very best scientists and fund them with no strings attached. 'I had 13 years of freedom at BP,' recalls Don Braben, who ran the programme and now serves in the office of vice-provost for research at University College London. 'We had 10,000 applicants and I picked just 37,' he says. 'Fourteen of those won major breakthroughs.'

That's a lesson funding agencies still need to heed, says Collins. 'It's difficult to have a policy to encourage serendipity,' he says. 'But it's not difficult to have a policy to discourage it.' Winning research grants is now so competitive – with just 10 per cent of applicants getting funded in many cases – that researchers have to play it safe and go after results they know they can achieve, he says. More adventurous proposals, those that might stumble across something altogether new, tend to be too risky to gain funding.

In essence, today's system is a self-fulfilling prophecy: it doesn't believe in chance and so chance discoveries seldom happen. Yet, with some enlightened thought – and a little bit of luck – that could be reversed.

Taking advantage of Lady Luck

The 19th-century chemist William Perkin was trying to synthesise the colourless antimalarial drug quinine from coal tar when he ended up with a vivid purple compound: the world's first synthetic organic dye.

Inspired by the burrs that stuck to his trousers after a hike, the inventor George de Mestral went on to develop Velcro.

Roy Plunkett, a chemist for DuPont, was working on a new chlorofluorocarbon refrigerant when he noticed that it left a slippery coating on its container. It's now widely used and goes by the name of Teflon.

In the 1930s, Karl Jansky, an engineer at Bell Labs, was investigating noise in transatlantic radio transmissions when he discovered that the static came from a fixed direction in the sky. This observation founded the field of radio astronomy.

Barnett Rosenberg was studying the effect of electricity on bacteria in the 1960s when he noticed that some of the cells had lost the ability to divide. The culprit turned out to be a by-product from a platinum electrode. We now know it as cisplatin, one of the most effective anticancer drugs.

Crunching the numbers

The strange mathematics of chance

What, exactly, is chance? Can we quantify it, or treat it like a set of biological specimens and classify it? Does it come in different strengths? This chapter is all about the science and mathematics of luck and coincidence, but it's far from just numbers – here you'll come face to face with hardened criminals, get the inside track on pharmaceutical trials, experience a supernova explosion and encounter Omega, the universe's most random number.

In the lap of the gods

How good are you at calculating the odds of anything? Are you a seasoned pattern-hound, or do you go crazy for coincidence? In the end, all this complexity is just about information – and who has it. Here's Ian Stewart.

As you'll know by now, the human brain is wonderful at spotting patterns. It's an ability that provides one of the foundation stones of science. When we notice a pattern, we try to pin it down mathematically, and then use the maths to help us understand the world around us. And if we can't spot a pattern, we don't put its absence down to ignorance. Instead we fall back on our favourite alternative. We call it randomness.

We see no patterns in the tossing of a coin, the rolling of dice, the spin of a roulette wheel, so we call them random. Until recently we saw no patterns in the weather, the onset of epidemics or the turbulent flow of a fluid, and we called them random too. It turns out that 'random' describes several different things: it may be inherent, or it may simply reflect human ignorance.

Little more than a century ago, it all seemed straightforward. Some natural phenomena were ruled by physical laws: the orbits of the planets, the rise and fall of the tides. Others were not: the pattern of hailstones on a path, for example. The first breach in the wall between order and chaos was the discovery, by Adolphe Quetelet around 1870, that there are statistical patterns in random events. The more recent discovery of chaos – apparently random behaviour in systems ruled by rigid laws – demolished parts of the wall completely. Whatever the ultimate resolution of order and chaos may be, they cannot be simple opposites.

Yet we still can't seem to resist the temptation of discussing real-world processes as if they are either ordered or random. Is the weather truly random or does it have aspects of pattern? Do dice really produce random numbers or are they in fact deterministic? Physicists have made randomness the absolute basis of quantum mechanics, the science of the very small: no one, they say, can predict when a radioactive atom will decay. But if that is true, what triggers the event? How does an atom 'know' when to decay? To answer these questions, we must sort out what kind of randomness we are talking about. Is it a genuine feature of reality or an artefact of how we model reality?

Let's start with the simplest ideas. A system can be said to be random if what it does next does not depend upon what it has done in the past. If I toss a 'fair' coin and get six heads in a row, the seventh toss can equally well be heads or tails. Conversely, a system is ordered if its past history affects its future in a predictable way. We can predict the next sunrise to within fractions of a second, and every morning we are right. So a coin is random but sunrise is not.

The pattern of sunrise stems from the regular geometry of the Earth's orbit. The statistical pattern of a random coin is more puzzling. Experiments show that in the long run, heads and tails turn up equally often, provided the coin is fair. If we think of the probability of an event as the proportion of times that it happens in a long series of trials, then both heads and tails have probability ½. That's not actually how probability is defined, but it is a simple consequence of the technical definition, called the law of large numbers.

The way coin tosses even out in the long run is a purely statistical feature of large numbers of tosses (see 'The law of averages', page 106). A deeper question, with a far more puzzling answer, is: how does the coin 'know' that it should be equally likely to come down heads as tails? The answer, when you look more closely, is that a coin is not a random system at all.

We can model the coin as a thin, circular disc. If the disc is launched vertically with a known speed and a known rate of rotation we can work out exactly how many half-turns it will make before it hits the floor and comes to rest. If it bounces, the calculation is harder but in principle it can be done. A tossed coin is a classical mechanical system. It obeys the same laws of motion and gravity that make the orbits of planets predictable. So why isn't the coin predictable?

Well, it is – in principle. In practice, however, you don't know the upward speed or the rate of spin, and it so happens that the outcome is very sensitive to both. From the moment you toss a coin – ignoring wind, a passing cat and other extraneous features – its fate is determined. But because you don't know the speed or the rate of spin, you have no idea what that inevitable fate is, even if you are incredibly quick at doing the sums.

A dice is the same. You can model it as a bouncing cube whose behaviour is mechanical and is governed by deterministic equations. If you could monitor the initial motion accurately enough and do the sums fast enough you could predict the exact result. Something along these lines has been done for roulette. The prediction is less precise – which half of the wheel the ball will end up in – but that's good enough to win, and the results don't have to be perfect to take the casino to the cleaners.

So when Albert Einstein questioned the randomness of quantum mechanics, refusing to believe that God throws dice, he chose entirely the wrong metaphor. He should have believed that God does play dice. Then he could have asked how the dice behave, where they are located, and what the real source of quantum 'randomness' is.

There is, however, a second layer to the problem. The difficulty in predicting the roll of a dice is not just caused by ignorance of the initial conditions. It is made worse by the curious nature of the process: it is chaotic.

Chaos is not random, but the limitations on the accuracy of any measurement we can make means it is unpredictable. In a random system, the past has no effect on the future. In a chaotic system, the past does have an effect on the future but the sums that ought to let us work out what the effect will be are extremely sensitive to tiny observational errors. Any initial error, however small, grows so rapidly that it ruins the prediction.

A tossed coin is a bit like that: a large enough error in measuring the initial speed and spin rate will stop us knowing the outcome. But a coin is not truly chaotic, because that error grows relatively slowly as the coin turns in the air. In a genuinely chaotic system, the error grows exponentially fast. The sharp corners of dice, which come into play when the perfect mathematical cube bounces off the flat table top, introduce this kind of exponential divergence. So dice seem random for two reasons: human ignorance of initial conditions as with the coin, and chaotic (though deterministic) dynamics.

Everything I have described so far has depended on the mathematical model that was chosen to describe it. So does the randomness, or not, of a given physical system depend on the model you use?

To answer that, let's take a look at the first great success of random models in physics: statistical mechanics. This theory underpins thermodynamics – the physics of gases – which was to some extent motivated by the need to make more efficient steam engines. How efficient can a steam engine get? Thermodynamics imposes very specific limits.

In the early days of thermodynamics, attention was directed at large-scale variables like volume, pressure, temperature and quantities of heat. The 'gas laws' connect these variables. For instance, Boyle's law says that the pressure of a sample of gas multiplied by its volume is constant at any given temperature. This is an entirely deterministic law: given the volume you can calculate the pressure, or vice versa.

However, it soon became apparent that the atomic-scale physics of gases, which underlies the gas laws, is effectively random: molecules of gas bounce erratically off each other. Ludwig Boltzmann was the first to explore how bouncing molecules, modelled as tiny hard spheres, relate to the gas laws (and much else). In his theory, the classical variables – pressure, volume and temperature – appeared as statistical averages that assumed an inherent randomness. Was this assumption justified?

Just as coins and dice are at root deterministic, so is a system composed of vast numbers of tiny hard spheres. It is cosmic snooker, and each ball obeys the laws of mechanics. If you know the initial position and velocity of every sphere, the subsequent motion is completely deter-mined. But instead of trying to follow the precise path of every sphere, Boltzmann assumed that the positions and speeds of the spheres have a statistical pattern that is not skewed in favour of any particular direction. Pressure, for example, is a measure of the average force exerted when the spheres bounce off the inner walls of their container, assuming that the spheres are equally likely to be travelling in any direction.

Statistical mechanics couches the deterministic motion of a large number of spheres in terms of statistical measures, such as an average. In other words, it uses a random model on the microscopic level to justify a deterministic model on the macroscopic level. Is that fair?

Yes it is, though Boltzmann didn't know it at the time. He effectively made two assertions: that the motion of the spheres is chaotic, and that the chaos is of a special kind that gives a well-defined average state. A whole branch of mathematics, ergodic theory, grew from these ideas, and the mathematics has advanced to the point where Boltzmann's hypothesis is now a theorem.

The change of viewpoint here is fascinating. An initially deterministic model (the gas laws) was refined to a random one (tiny spheres), and the randomness was then justified mathematically as a consequence of deterministic dynamics.

So are gases really random or not? It all depends on your point of view. Some aspects are best modelled statistically, others are best modelled deterministically. There is no one answer, it depends on the context. This situation is not at all unusual. For some purposes – calculating the airflow over the space shuttle, for example – a fluid can be considered as a continuum, obeying deterministic laws. For other purposes, such as Brownian motion – the erratic movement of suspended particles caused by atoms bouncing into them – the atomic nature of the fluid must be taken into account and a Boltzmann-like model is appropriate.

So we have two different models with a mathematical link between them. Neither is reality, but both describe it well. And it doesn't seem to make any sense to say that the reality is or is not random: randomness is a mathematical feature of how we think about the system, not a feature of the system itself.

So is nothing truly random? Until we understand the roots of the quantum world, we can't say for sure. In its usual interpretations, quantum mechanics asserts that deep down, on the subatomic level, the universe is genuinely and irreducibly random. It is not like the hard-spheres model of thermodynamic randomness, which traces the statistical features to our (unavoidable) ignorance of the precise state of all the spheres. There is no analogous small-scale model with a few parameters that, if we could only see them, would unlock the mystery. The 'hidden variables', whose deterministic but chaotic behaviour governs the throw of the quantum dice, simply don't exist. Quantum stuff is random, period. Or is it?

There is certainly a mathematical argument to justify such an assertion. In 1964 John Bell came up with a way of testing whether quantum mechanics is random or governed by hidden variables – essentially, quantum properties that we have not yet learned how to observe. Bell's work was centred on the idea of two quantum particles, such as electrons, that interact and are then separated over vast distances. Perform a particular set of measurements on these widely separated particles and you should be

able to determine whether their properties are under-
pinned by randomness or in thrall to hidden variables.
The answer is important: it dictates whether quantum
systems that have interacted in the past are subsequently
able to influence each other's properties – even if they are
at opposite ends of the universe.

As far as most physicists are concerned, experiments
based on Bell's work have confirmed that, in quantum
systems, randomness – and the bizarre 'action at a distance' – rules. Indeed, so keen are they to put over the
fundamental role of randomness in quantum theory that
they tend to dismiss any attempt to question it further.
This is a pity, because Bell's work, though brilliant, is not
as conclusive as they imagine.

The issues are complex, but the basic point is that mathematical theorems involve assumptions. Bell makes his
main assumptions explicit, but the proof of his theorem
involves some implicit assumptions too, something that
is not widely recognised. There are also open loopholes in
the experimental versions of Bell's work. These are largely
technical – to do with detector efficiencies and experimental errors – but there are philosophical aspects too: the
experiments assume, for instance, that the human beings
doing the experiment are free to choose its parameters. It
is possible, though seemingly unlikely, that an external
force could be co-ordinating and controlling every aspect
of the experiment – including the experimenters.

So, despite the vast weight of opinion, the door is still
open for a deterministic explanation of quantum indeter-
minacy. The devil, as always, is in the detail. It may be
difficult, or even impossible, to test such a theory, but we

can't know that until someone writes it down. It may not change quantum mechanics much, any more than hard spheres changed thermodynamics, but it would give us an entirely new insight into many puzzling questions. And it would put quantum theory back among all the other statistical theories of science: random from some points of view, deterministic from others.

In the meantime, quantum stuff apart, we can state with assurance that there really is no such thing as randomness. Virtually all apparently random effects arise not because nature is genuinely unpredictable but because of human ignorance or other limitations on possible knowledge of the world. This insight is not new. Alexander Pope, in his *Essay on Man*, wrote: 'All nature is but art, unknown to thee/ All chance, direction which thou canst not see/ All discord, harmony not understood/ All partial evil, universal good.' Apart from the bit about good and evil, mathematicians now understand precisely why he was right.

The law of averages

I once threw 17 consecutive heads with a normal coin, an event with probability 1 in 131,072. Surely tails must now need to 'catch up', and be more likely as the result of the next throw?

Not so. The next toss is just as likely to produce another head as a tail, and the same goes for all subsequent tosses. In the long run, subsequent tosses should be very close to half heads, half tails. So, in 2 million additional tosses, we expect, on average, a million heads and a million tails.

Although 17 is very different from zero, 1,000,017 is

proportionately much closer to a million: their ratio is 1.000017, very close to 1. Instead of tails catching up with heads, the future tosses swamp the first few, and the longer you keep tossing the less important that initial difference becomes.

It's related to how frequently numbers appear in the UK's Lotto draws. At one stage 13 was relatively infrequent, reinforcing the view that 13 is unlucky. Some people therefore expect 13 to come up more often in future. Others think that its unluckiness will persist. The mathematics of probability, supported by innumerable experiments, says both camps are wrong. In future, all numbers have the same chance of being picked. The lottery machine treats all balls alike, and it doesn't 'know' what number is written on them.

Paradoxically, that does not mean every number will turn up equally often. Exact equality is highly unlikely. Instead, we expect to see fluctuations around the average value, with some winners and some losers. The mathematics even predicts the size and likelihood of those fluctuations. What the maths can't do is predict which numbers will be winners and which losers. In advance, it could equally well be any of them.

Is that supposed to happen?

It's all very well considering abstract notions of randomness, but everyday life has a way of constraining what should be chance occurrence. Cross random chance with the real world, and there are strange and wonderful mathematical consequences that have a deep resonance with natural phenomena. It's almost as if randomness can lose itself, says Robert Matthews.

Many people would call them spooky: bizarre coincidences that loom out at us from the randomness of everyday events. But everyone knows that randomness is the very essence of patternless, lawless disorder. It's obvious that there can be nothing in these coincidences.

Obvious, but wrong. Peer hard enough into the fog of randomness, and you can glimpse regularities and universal truths normally associated with deep cosmic order. Why? Because what we call randomness is typically only a chained and muzzled version of the real thing. When forced to act within certain limits, imposed on it by the constraints of the world we live in, randomness sheds just a little of its notorious mathematical lawlessness. The effects are usually subtle, but sometimes shocking and as plain as day – when you know where to look.

Take lottery numbers, for example. A quick look at past winning combinations suggests nothing but randomness. But look harder, and tiny specks of order start to appear: a pair of consecutive numbers here, a run of prime numbers there.

But no one is fixing the lottery – statistical tests are carried out to keep a watch for this – so what is going on? What we're seeing is the revenge of randomness, its retaliation for being hobbled. Truly random numbers know no bounds but those in the lottery have no such freedom. They're confined to a range – like 1 to 49. And whenever randomness has its style cramped in this way, with only certain outcomes allowed, it loses some of its utter lawlessness and unpredictability. Instead, it must fall into line with probability theory, which describes the behaviour of infinite randomness in a finite world.

In the case of a 49-ball lottery, for example, probability theory proves that examples of apparently anomalous order will show up in roughly half of all draws. When randomness is compelled to scatter surprises among just a limited number of outcomes, we should expect the unexpected.

Take a random weekend of the football season in any year in any country – say, 14/15 August of the English Premier League's 2004 season. Twenty teams played each other that weekend in ten matches, and half of those matches featured players sharing the same birthday. A bizarre coincidence? No. In fact, probability theory shows that when randomness is forced to scatter the birthdays of the 22 players in each match among the 365 days of a year, there is a roughly 50:50 chance that at least two players in a match will share the same birthday. In other words, around half of the ten matches played on that first weekend should have seen at least two players sharing the same birthday. And that's exactly what happened.

Probability theory also predicted a roughly 50:50 chance that at least one player out of the 230 playing that weekend would be celebrating his birthday on the day of the match. In fact, two were: Jay-Jay Okocha of Bolton Wanderers and Johnnie Jackson of Tottenham Hotspur.

Looking harder still at randomness reveals more subtle signs of its revolt against constraint. Around a century ago, the statistician Ladislaus Bortkiewicz produced a classic study of fatalities in the Prussian army that highlights a bizarre link between randomness and a universal mathematical constant known as e. This never-ending decimal number, which begins 2.718281..., often pops up in natural processes where the rate of some process depends on the present state of the system, such as the rate of growth of populations, or radioactive decay.

Bortkiewicz's data shows this universal number can also be found lurking in random events, such as the risk of death from a horse-kick. According to the reports, the Prussian soldiers all faced a small but finite risk of death from horse-kicks, amounting to an average of one fatality every 1.64 years. Bortkiewicz found that of the 200 reports, 109 recorded no deaths at all. Now divide 200 by 109, and raise the result to the power 1.64, the average interval between deaths through horse-kicks. The result is 2.71 – within 1 per cent of e.

A fluke? Not at all: it's to do with the mathematics of what are called Poisson distributions. Probability theory shows that e can be expected to pop up when lots of randomly triggered events are spread over a restricted interval of time. The same is true of events spread over a limited region of space: you can extract a value of e from the impact sites of the V-1 'flying bombs' targeted on south London during the Second World War. While there were hundreds of impacts, the chances of randomness landing a V-1 on a specific part of the capital were low. And analysing the data in the same way as for horse-kick deaths leads to a value for e of 2.69 – again, within around 1 per cent of the exact value.

It's a similar story with the rate at which wars break out between nations over the years, and many other human phenomena. In each case, the chances of the event may be low, but there are lots of opportunities for it to happen, and randomness responds by allowing e to inveigle its way into the data.

Suitably cajoled, randomness will also produce values for probably the most famous universal constant of all. Drop a needle carelessly onto wooden floorboards: the number of times it falls across a gap between the floorboards depends on the dimensions of the needle, the floorboards and ... π. It appears because of the random angle at which the needle ends up on the floor. Observe a few tens of thousands of such events and a surprisingly accurate value of π emerges from the randomness.

It's possible to pluck the white rabbit of π from any hat brimming with randomness. Gather a hefty number of random integers and check whether each pair has any common factor. Take the proportion of the total that don't, multiply it by 6 and then take the square root: a mathematical theorem shows that as the number of pairs examined increases, that little recipe will give a number that gets ever closer to π.

You can even use the stars in the night sky: all you need to do is compare the angular distance between any two stars on the celestial sphere with that of any other pair. Do this for the 100 brightest stars in the sky, and the common-factor method gives you a 'celestial' value for π of 3.12772 – within 0.5 per cent of the true value.

We humans seem to have a penchant for seeing patterns in randomness, from religious figures in clouds to faces on Mars. We're right to dismiss most of them as nothing more than illusions. But sometimes randomness can spring surprises on us, with patterns that hint at the order lurking behind everything.

Rough justice

If you want to understand the chances of something – the DNA found at a crime scene matching the accused's, for example – you need a good grounding in statistics. But what constitutes a good grounding has changed in recent years. For many people, traditional statistics needs updating with an 18th-century idea, as Angela Saini explains.

Shambling sleuth Columbo always gets his man. Take, for example, the society photographer in a 1974 episode of the cult US television series. He has killed his wife and disguised it as a bungled kidnapping. It is the perfect crime – until the hangdog detective hits on a cunning ruse to expose it. Columbo induces the murderer to grab from a shelf of 12 cameras the exact one used to snap the victim before she was killed. 'You just incriminated yourself, sir,' says a watching police officer.

If only it were that simple. Killer or not, anyone would have a 1 in 12 chance of picking the same camera at random. That kind of evidence would never stand up in court. Or would it?

Such probabilistic pitfalls are not limited to crime fiction. 'Statistical errors happen astonishingly often,' says Ray Hill, a mathematician at the University of Salford, UK, who has given evidence in several high-profile criminal cases. 'I'm always finding examples that go unnoticed in evidence statements.'

The root cause is a sloppiness in analysing odds that can sully justice and even land innocent people in jail. With ever more trials resting on the 'certainties' of data such as DNA matches, the problem is becoming more acute. Some mathematicians are calling for the courts to take a crash course in the true significance of the evidence put before them. Their demand? Bayesian justice for all.

That rallying call derives from the work of Thomas Bayes, an 18th-century English mathematician and clergyman who showed how to calculate conditional probability – the chance of something being true if its truth depends on other things also being true. That is precisely the kind of problem that criminal trials deal with as they sift through evidence to establish a defendant's innocence or guilt (see 'Bayes on trial', page 114).

Mathematics might seem a logical fit for the courts, then. Judges and juries, though, rely on gut feeling all too often. A startling example was the rape trial in 1996 of a British man, Dennis John Adams. Adams hadn't been identified in a line-up and his girlfriend had provided an alibi. But his DNA was a 1 in 200 million match to semen from the crime scene – evidence seemingly so damning that any jury would be likely to convict him.

The issue we have to face, though, is what that figure actually meant. It is not, as courts and the press often assume, that there is only a 1 in 200 million chance that the semen belonged to someone other than Adams, making his innocence implausible. As we will discover later on, it actually means there is a 1 in 200 million chance that the DNA of any random member of the public will match that found at the crime scene.

The difference is subtle, but significant. In a population, say, of 10,000 men who could have committed the crime, there would be a 10,000 in 200 million, or 1 in 20,000,

Bayes on trial

Could you be a Bayesian juror? As the following example shows, it's not straightforward. Suppose you have a piece of evidence, E, from a crime scene – a bloodstain, or perhaps a clothing thread – that matches to a suspect. How should it affect your perception or hypothesis, H, of the suspect's innocence?

Bayes's theorem tells you how to work out the probability of H given E. It is: (the probability of H) multiplied by (the probability of E given H) divided by (the probability of E). Or in standard mathematical notation:

$P(H \mid E) = P(H) \times P(E \mid H)/P(E)$

Say you are a juror at an assault trial, and so far you are 60 per cent convinced the defendant is innocent: $P(H) = 0.6$. Then you're told that the blood of the defendant and blood found at the crime scene are both type B, which is found in about 10 per cent of people. How should this change your view? Is their guilt more or less likely?

What the forensics expert has given you is the probability that the evidence matches anyone in the general population, given that they are innocent: $P(E \mid H) = 0.1$. To apply Bayes's formula and find $P(H \mid E)$ – your new estimation of the defendant's innocence – you now need the quantity $P(E)$, the probability that their blood matches that at the crime scene.

This probability actually depends on the defendant's innocence or guilt. If they are innocent, it is 0.1 as it is for anyone else. If they are guilty, however, it is 1, as their blood is certain to match. This insight allows us to calculate $P(E)$ by summing the probabilities of a blood match in the case of innocence (H) or guilt (not H):

$$P(E) = [P(E \mid H) \times P(H)] + [P(E \mid \text{not } H) \times P(\text{not } H)] = (0.1 \times 0.6) + (1 \times 0.4) = 0.46$$

So according to Bayes's formula the revised probability of their innocence is:

$$P(H \mid E) = (0.6 \times 0.1) \div 0.46 = 0.13$$

As you might expect, by this measure the probability of innocence has tumbled. The defendant is between four and five times guiltier than you first thought – probably.

chance that someone else is a match too. That still doesn't look good for Adams, but it's not nearly as damning.

So worried was Adams's defence team that the jury might misinterpret the odds that they called in Peter Donnelly, a statistical scientist at the University of Oxford. 'We designed a questionnaire to help them combine all the evidence using Bayesian reasoning,' says Donnelly.

They failed to convince the jury of the value of the Bayesian approach, however, and Adams was convicted. He appealed twice unsuccessfully, with an appeal judge eventually ruling that the jury's job was 'to evaluate evidence not by means of a formula ... but by the joint application of their individual common sense.'

But what if common sense runs counter to justice? For David Lucy, a mathematician at Lancaster University in the UK, the Adams judgement indicates a cultural tradition that needs changing. 'In some cases, statistical analysis is the only way to evaluate evidence, because intuition can lead to outcomes based upon fallacies,' he says.

Norman Fenton, a computer scientist at Queen Mary University of London who has worked for defence teams in criminal trials, has come up with a possible solution. With his colleague Martin Neil, he developed a system of step-by-step pictures and decision trees to help jurors grasp Bayesian reasoning. Once a jury has been convinced that the method works, the duo argue, experts should be allowed to apply Bayes's theorem to the facts of the case as a kind of 'black box' that calculates how the probability of innocence or guilt changes as each piece of evidence is presented. 'You wouldn't question the steps of an electronic calculator, so why here?' Fenton asks.

It is a controversial suggestion. Taken to its logical conclusion, it might see the outcome of a trial balance on a single calculation. Working out Bayesian probabilities with DNA and blood matches is all very well, but quantifying incriminating factors such as appearance and behaviour is more difficult. 'Different jurors will interpret different bits of evidence differently. It's not the job of a mathematician to do it for them,' Donnelly says.

He thinks forensics experts should be schooled in statistics so that they can catch errors before they occur. And since cases such as that of Adams, this has indeed begun to happen in the US and UK. Lawyers and jurors, however, still have far less – if any – statistical training.

As the five real-life fallacies that follow show, there's no room for complacency. According to Donnelly, these examples from the legal casebook demonstrate that the call for a revision in statistical analysis is not about mathematicians trying to force their way of thinking on the world. 'Justice depends on getting everyone to reason properly with uncertainties,' he says.

1) Prosecutor's fallacy

'The prosecutor's fallacy is such an easy mistake to make,' says Ian Evett of Principal Forensic Services, an evidence consultancy based in Kent, England. It confuses two subtly different probabilities that Bayes's formula distinguishes: $P(H \mid E)$, the probability that someone is innocent if they are a match to a piece of evidence, and $P(E \mid H)$, the probability that someone is a match to a piece of evidence if they are innocent. The first probability is what we would like to know; the second is what forensics usually tells us.

Unfortunately, even professionals sometimes mix them up. In the 1991 rape trial of Andrew Deen in Manchester, UK, for example, an expert witness agreed on the basis of a DNA sample that 'the likelihood of [the source of the semen] being any other man but Andrew Deen [is] 1 in 3 million.'

That was wrong. One in 3 million was the likelihood that any innocent person in the general population had a DNA profile matching that extracted from semen at the crime scene – in other words, $P(E \mid H)$. With around 60 million people in the UK, a fair few people will share that profile. Depending on how many of them might plausibly have committed the crime, the probability of Deen being innocent even though he was a match, or $P(H \mid E)$, was actually far greater than 1 in 3 million.

Deen's conviction was quashed on appeal, leading to a flurry of similar appeals that have had varying success – and some surprise outcomes, such as a Californian man jailed in 2008, who was discovered by police to be a DNA match to a rape and murder committed 35 years earlier.

2) Ultimate issue error

The prosecution in the Deen case stopped just short of compounding their probabilistic fallacy. In the minds of the jury, though, it probably morphed into the 'ultimate issue' error: explicitly equating the (small) number P(E | H) with a suspect's likelihood of innocence.

In Los Angeles in 1968, the ultimate issue error sent Malcolm Collins and his wife Janet to jail. At first glance, the circumstances of the case left little room for doubt: an elderly lady had been robbed by a white woman with blonde hair and a black man with a moustache, who had both fled in a yellow car. The chances of finding a similar interracial couple matching that description were 1 in 12 million, an expert calculated.

The police were convinced, and without much deliberation so was the jury. They assumed that there was a 1 in 12 million chance that the couple were not the match, and that this was also the likelihood of their innocence.

They were wrong on both counts. In a city such as Los Angeles, with millions of people of all races living in it or passing through, there could well have been at least one other such couple, giving the Collinses an evens or better chance of being innocent. Not to mention that the description itself might have been inaccurate – facts that helped reverse the guilty verdict on appeal.

3) Base-rate neglect

Anyone looking to DNA profiling for a quick route to a conviction should recognise that genetic evidence can be shaky. Even if the odds of finding another genetic match are 1 in a billion, in a world of 7 billion, that's another seven people with the same profile.

Fortunately, circumstantial and forensic evidence often quickly whittle down the pool of suspects. But neglecting your 'base rate' – the pool of possible matches – can have you leap to false conclusions, not just in the courtroom.

Picture yourself, for example, in the doctor's surgery. You have just tested positive for a terminal disease that afflicts 1 in 10,000. The test has an accuracy of 99 per cent. What's the probability that you actually have the disease?

It is in fact less than 1 per cent. The reason is the sheer rarity of the disease, which means that even with a 99 per cent accurate test, false positives will far outweigh real ones (see diagram). That's why it is so important to carry out further tests to narrow down the odds. We lay people are not the only ones stumped by such counter-intuitive results: surveys show that 85 to 90 per cent of health professionals get it wrong too.

4) Defendant's fallacy

It's not just prosecutors who can fiddle courtroom statistics to their advantage: defence lawyers have also been known to cherry-pick probabilities.

In 1995, for example, former American football star O. J. Simpson stood trial for the murder of his ex-wife, Nicole Brown, and her friend. Years before, Simpson had pleaded no contest to a charge of domestic violence against Brown. In an attempt to downplay that, a consultant to Simpson's defence team, Alan Dershowitz, stated that fewer than 1 in 1,000 women who are abused by their husbands or boyfriends end up murdered by them.

That might well be true, but it was not the most relevant fact, as John Allen Paulos, a mathematician at Temple University in Philadelphia, Pennsylvania, later showed. As a Bayesian calculation taking in all the pertinent facts reveals, it is trumped by the 80 per cent likelihood that, if a woman is abused and later murdered, the culprit was her partner.

That may not be the whole story either, says criminologist William Thompson of the University of California, Irvine. If more than 80 per cent of all murdered women, abused or not, are killed by their partner, 'the presence of abuse may have no diagnostic value at all'.

5) Dependent evidence fallacy

Sometimes, mathematical logic flies out of the courtroom window long before Bayes can even be applied – because the probabilities used are wrong.

Take the dependent evidence fallacy, which was central to one of the most notorious recent miscarriages of justice in the UK. In November 1999, Sally Clark was convicted of smothering her two children as they slept. A paediatrician, Roy Meadow, testified that the odds of both dying naturally by sudden infant death syndrome (SIDS), or cot death, were 1 in 73 million. He arrived at this figure by multiplying the individual probability of SIDS in a family such as Clark's – 1 in 8,500 – by itself, as if the two deaths were independent events.

Don't panic

You've just been diagnosed with a rare condition that afflicts
1 in 10,000. The test is 99 per cent certain. Hope or despair?

■ True positive ■ False positive

In a population of 10,000, on average one person will have the disease –
and they will also test positive

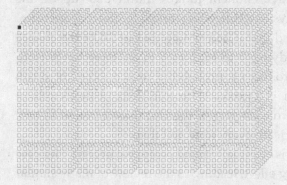

If the test is only 99 per cent accurate, 1 per cent of the remaining
healthy population will test positive too

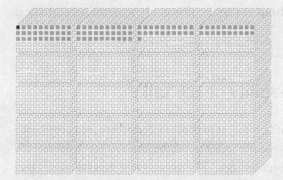

So if you test positive, all other things being equal, there's a chance of
over 99 per cent that you **don't** have the disease – **HOPE**

But why should they be? 'There may well be unknown genetic or environmental factors that predispose families to SIDS, so that a second case within the family becomes much more likely,' the Royal Statistical Society explained during an appeal.

'Even three eminent judges didn't pick up on the mistake,' says Ray Hill, who worked for the defence team. He estimated that if one sibling dies of SIDS, the chance of another dying is as high as 1 in 60. Bayesian reasoning then produces a probability of a double cot death of around 1 in 130,000. With hundreds of thousands of children born each year in the UK, there's bound to be a double cot death every now and then.

Clark was eventually freed on appeal in 2003. Her case had a lasting effect, leading to the review of many similar cases. 'I'm not aware of any cases of multiple cot deaths reaching the courts in recent years,' says Hill. Clark herself never recovered from her ordeal, however. She was found dead at her home in 2007, ultimately a victim of statistical ignorance.

The probability peace talks

Given that we've brought back an 18th-century idea about statistics, what is the likelihood it will become our favoured method for divining probabilities? The answer will depend on whether we can combine two very different worlds, as Regina Nuzzo explains.

Let's start with an old T-shirt slogan: 'Statistics means

never having to say you're certain.' Drawing conclusions without all the facts is the bread-and-butter of statistics. How many people in a country support legalising cannabis? You can't ask all of them. Is a run of hotter summers consistent with natural variability, or a trend? There's no way to look into the future to say definitively.

Answers to such questions generally come with a probability attached. But that single number often masks a crucial distinction between two different sorts of uncertainty: stuff we don't know, and stuff we can't know.

Can't-know uncertainty results from real-world processes whose outcomes appear random to all who look at them: how a dice rolls, where a roulette wheel stops, when exactly an atom in a radioactive sample will decay. This is the world of 'frequentist' probability, because if you roll enough dice or observe enough atoms decaying, you can get a reasonable handle on the relative frequency of different outcomes, and can construct a measure of their probabilities.

Don't-know uncertainty is more slippery. Here individual ignorance, not simply universal randomness, is at play. What's the sex of your pregnant neighbour's baby? It's already a given, so there's no chance involved – but you don't yet know, so you are uncertain. If you are into in-play gambling, you might wonder: who will win an ongoing soccer match? This isn't purely random, either. If you're paying close attention to the game's story so far, you are more certain about the outcome than someone who has been napping throughout. Welcome to Bayesian statistics.

How to approach these different types of uncertainty

divides frequentists and Bayesians. A strict frequentist has no truck with don't-know uncertainty, or any probability measure that can't be derived from repeatable experiments, random-number generators, surveys of a random population sample and the like. A Bayesian, meanwhile, doesn't bat an eyelid at using other 'priors' – knowledge gleaned from past voting patterns in a general election, for example – to fill in the gaps. 'Bayesians are happy to put probabilities on statements about the world – frequentists aren't,' says Tony O'Hagan, a statistician at the University of Sheffield, UK, who researches Bayesian methods. 'In the Bayesian approach we try to answer questions by bringing all the relevant evidence to bear on it, even when the contribution of some of that evidence to the question depends on subjective judgements,' he says.

In the late 18th and early 19th centuries, Bayesian-style methods helped tame a range of inscrutable problems, from estimating the mass of Jupiter to calculating the number of boys born worldwide for every girl. But it gradually fell out of favour, victim of a dawning era of big data. Everything from improved astronomical observations to newly published statistical tables of mortality, disease and crime conveyed a reassuring air of objectivity. Bayes's methods of educated guesswork seem hopelessly old-fashioned, and rather unscientific by contrast. Frequentism, with its emphasis on dispassionate number crunching of the results of randomised experiments, came increasingly into vogue.

The advent of quantum theory in the early 20th century, which re-expressed even reality in the language of frequentist probability, provided a further spur to that development. The two strands of thought in statistics gradually drifted further apart. Adherents ended up submitting work to their own journals, attending their own conferences and even forming their own university departments. Emotions often ran high. The author Sharon Bertsch McGrayne recalls that when she started researching her book on the history of Bayesian ideas, *The Theory That Would Not Die*, one frequentist-leaning statistician berated her down the phone for attempting to legitimise Bayesianism. In return, Bayesians developed a sort of persecution complex, says Robert Kass at Carnegie Mellon. 'Some Bayesians got very self-righteous, with a kind of religious zealotry.'

In truth, though, both methods have their strengths and weaknesses. Where data points are scant and there is little chance of repeating an experiment, Bayesian methods can excel in squeezing out information. Take astrophysics as an example. A supernova explosion in a nearby galaxy, the Large Magellanic Cloud, seen in 1987, provided a chance to test long-held theories about the flux of neutrinos from such an event – but detectors picked up only 24 of these slippery particles. Without a profusion of repeatable data, frequentist methods fell down – but the flexible, information-borrowing Bayesian approach provided an ideal way to assess the merits of different competing theories.

It helped that well-grounded theories provided good priors to start that analysis. Where these don't exist, a Bayesian analysis can easily be a case of garbage in, garbage out. It's one reason why courts of law have been wary of adopting Bayesian methods, even though on the face of it they are an ideal way to synthesise messy evidence from many sources. In a 1993 New Jersey paternity case that used Bayesian statistics, the court decided jurors should use their own individual priors for the likelihood of the defendant having fathered the child, even though this would give each juror a different final statistical estimate of guilt. 'There's no such thing as a right or wrong Bayesian answer,' says Larry Wasserman of Carnegie Mellon University. 'It's very postmodern.'

Finding good priors can also demand an impossible depth of knowledge. Researchers searching for a cause for Alzheimer's disease, for instance, might test 5,000 genes. Bayesian methods would mean providing 5,000 priors for the likely contribution of each gene, plus another 25 million if they wanted to look for pairs of genes working together. 'No one could construct a reasonable prior for such a high-dimensional problem,' says Wasserman. 'And even if they did, no one else would believe it.'

To be fair, without any background information, standard frequentist methods of sifting through many tiny genetic effects wouldn't easily allow the truly important genes and combinations of genes to rise to the top of the pile. But this is perhaps a problem more easily dealt with than conjuring up 25 million coherent Bayesian guesses.

Frequentism in general works well where plentiful data should speak in the most objective way possible. One high-profile example is the search for the Higgs boson, completed in 2012 at the CERN particle physics laboratory near Geneva, Switzerland. The analysis teams concluded that if in fact there were no Higgs boson, then a pattern of data as surprising as, or more surprising than, what was observed would be expected in only one in 3.5 million hypothetical repeated trials. That is so unlikely that the team felt comfortable rejecting the idea of a universe without a Higgs boson.

That wording may seem convoluted to you. This highlights frequentism's main weakness: the way it ties itself in knots through its disdain for all don't-know uncertainties. The Higgs boson either exists or it doesn't, and any inability to say one way or the other is purely down to lack of information. A strict frequentist can't actually make a direct statement of the probability of its existing or not – as indeed the CERN researchers were careful not to (although certain sections of the media and others were freer).

Head-to-head comparisons can point to the confusions that can arise from this, as was the case with a controversial clinical trial of two heart-attack drugs, streptokinase and tissue plasminogen activator, in the 1990s. The first, frequentist analysis gave a 'p value' of 0.001 to a study seeming to show that more patients survived after the newer, more expensive tissue plasminogen activator therapy. This equates to saying that if the two drugs had the same mortality rate, then data at least as extreme as the observed rates would occur only once in every 1,000 repeated trials.

That doesn't mean the researchers were 99.9 per cent certain the new drug was superior – although again it is often interpreted that way. When other researchers conducted a Bayesian reanalysis using the results of previous clinical trials as a prior, they found a direct probability of the new drug being superior of only about 17 per cent.

'In Bayesianism we're directly addressing the question of interest, talking about how likely it is to be true,' says David Spiegelhalter of the University of Cambridge. 'Who wouldn't want to talk about that?'

Perhaps it's just a case of horses for courses, but don't the strengths and weaknesses of each approach suggest we might be better off combining elements of both? Kass is one of a new breed of statisticians doing just that. 'To me statistics is like a language,' he says. 'You can be conversant in both French and English and switch back and forth comfortably.'

Stephen Senn, a drugs statistician at the Luxembourg Institute of Health, agrees. 'I use what I call "mongrel statistics", a little bit of everything,' he says. 'I often work in a frequentist mode, but I reserve the right to do Bayesian analyses and think in a Bayesian way.'

Kass points to an analysis he and his colleagues did on the firing rates of a couple of hundred neurons in the visual-motor region of the brain in monkeys. Prior work in basic neurobiology provided them with information on how fast these neurons should be firing, and how quickly the rate might change over time. They incorporated this into a Bayesian approach, then switched gears to evaluate their results under a standard frequentist framework. The Bayesian prior gave the methods enough of a kick-start to allow frequentist methods to detect even tiny differences in a sea of noisy data. The two approaches together trumped either method alone.

Sometimes, Bayesian and frequentist ideas can be blended so much that they create something new. In large genomics studies, a Bayesian analysis might exploit the fact that a study testing the effects of 2,000 genes is almost like 2,000 parallel experiments, and so it can cross-fertilise the analyses, using the results from some to establish priors for others and using that to hone the conclusions of a frequentist analysis. 'This approach gives quite a bit better results,' says Jeff Leek of Johns Hopkins University in Baltimore, Maryland. 'It's really transformed the way we analyse genomic data.'

It breaks down barriers too. 'Is this approach frequentist? Bayesian?' asked Harvard University biostatistician Rafael Irizarry in a blog post. 'To this applied statistician, it doesn't really matter.'

Not that the arguments have entirely gone away. 'Statistics is essentially the abstract language that science uses on top of data to tell stories about how nature works, and there is not one unique way to tell stories,' says Kass. 'Two hundred years from now there might be some breakthrough connecting Bayesianism and frequentism into a grand synthesis, but my guess is that there will always be at least one versus the other.'

Known unknowns

In our final look at the mathematics of chance, we are going to head to the dark heart of number theory. Here be monsters: entirely unpredictable numbers that mean we cannot prove certain theorems. They were discovered by Gregory Chaitin, who lays out here exactly why some find the discovery so disturbing.

The question of predictability has a long history in physics.

In the early 19th century, the classical deterministic laws of Isaac Newton led Pierre Simon de Laplace to believe that the future of the universe could be determined for ever.

But then quantum mechanics came along. This is the theory that is fundamental to our understanding of the nature of matter, describing very small objects, such as electrons and other fundamental particles. One of the controversial features of quantum mechanics was that it introduced probability and randomness at a fundamental level to physics. This deeply upset the great physicist Albert Einstein, who said that God did not play dice.

A few decades later, the study of non-linear dynamics surprised us by showing that even the classical physics of Newton had randomness and unpredictability at its core. Randomness and unpredictability was beginning to look like a unifying principle.

And it seems that the same principle even extends to mathematics. I can show that there are theorems connected with number theory that cannot be proved because when we ask the appropriate questions, we obtain results that are equivalent to the random toss of a coin.

My results would have shocked many 19th-century mathematicians, who believed that mathematical truths could always be proved. For example, in 1900, the mathematician David Hilbert gave a famous lecture in which he proposed a list of 23 problems as a challenge to the new century. His sixth problem had to do with establishing the fundamental universal truths, or axioms, of physics. One of the points in this question concerned probability theory. To Hilbert, probability was simply a practical tool that came from physics; it helped to describe the real world when there was only a limited amount of information available.

Another question he discussed was his tenth problem, which was connected with solving 'diophantine' equations, named after the Greek mathematician Diophantus. These are algebraic equations involving only whole numbers, or integers. Hilbert asked: 'Is there a way of deciding whether or not an algebraic equation has a solution in whole numbers?'

Little did Hilbert imagine that the sixth and the tenth problems were subtly related. This was because he had assumed something that was so basic to his thinking that he did not even formulate it as a question in his talk – the idea that every mathematical problem has a solution. We may not be bright enough or we may not have worked long enough on a problem but, in principle, it should be possible to solve it – or so Hilbert thought. For him, it was a black-or-white situation.

It seems now that Hilbert was on shaky ground. In fact, there is a connection between Hilbert's sixth question dealing with probability theory and his tenth problem of solving algebraic equations in whole numbers. And this connection leads to a surprising and rather chilling result. That is: randomness lurks at the heart of that most traditional branch of pure mathematics, number theory.

It turns out that clear, simple mathematical questions do not always have clear answers. In elementary number theory, questions involving diophantine equations can give answers that are completely random and look grey, rather than black or white. The answer is random because the only way to prove it is to postulate each answer as an additional independent axiom. Einstein would be horrified to discover that not only does God play dice in quantum and classical physics but also in pure mathematics.

Where does this surprising conclusion come from? We have to go back to Hilbert. He said that when you set up a formal system of axioms there should be a mechanical procedure to decide whether a mathematical proof is correct or not, and the axioms should be consistent and complete. If the system of axioms is consistent, it means that you cannot prove both a result and its contrary. If the system is complete, then you can also prove any assertion to be true or false. It follows that a mechanical procedure would ensure that all mathematical assertions can be decided mechanically.

There is a colourful way to explain how this mechanical procedure works: the so-called 'British Museum algorithm'. What you do – it cannot be done in practice because it would take for ever – is to use the axiom system, set in the formal language of mathematics, to run through all possible proofs, in order of their size and lexicographic order. You check which proofs are correct – which ones follow the rules and are accepted as valid. In principle, if the set of axioms is consistent and complete, you can decide whether any theorem is true or false. Such a procedure means that a mathematician no longer needs ingenuity or inspiration to prove theorems. Mathematics becomes mechanical.

Of course, mathematics is not like that. Kurt Gödel, the Austrian logician, and Alan Turing, the father of the computer, showed that it is impossible to obtain both a consistent and complete axiomatic theory of mathematics and a mechanical procedure for deciding whether an arbitrary mathematical assertion is true or false, or is provable or not.

Gödel was the first to devise the ingenious proof, couched in number theory, of what is called the incompleteness theorem. But I think that the Turing version of the theorem is more fundamental and easier to understand. Turing used the language of the computer – the instructions, or program that a computer needs to work out problems – to show that there is no mechanical procedure for deciding whether an arbitrary program will ever finish its computation and halt.

To show that the so-called halting problem can never be solved, we set the program running on a Turing machine, which is a mathematical idealisation of a digital computer with no time limit. (The program must be self-contained with all its data wrapped up inside the program.) Then we simply ask: 'Will the program go on for ever, or at some point will it say "I'm finished" and halt?'

Turing showed that there is no set of instructions that you can give the computer, no algorithm, that will decide in advance if a given program will ever halt. Gödel's incompleteness theorem follows immediately because if there is no mechanical procedure for deciding the halting problem, then there is no complete set of underlying axioms either. If there were, they would provide a mechanical procedure for running through all possible proofs to show whether programs halt – although it would take a long time, of course.

To obtain my result about randomness in mathematics,

I simply take Turing's result and change the wording. What I get is a sort of mathematical pun. Although the halting problem is unsolvable, we can look at the probability of whether a randomly chosen program will halt. We start with a thought experiment using a general purpose computer that, given enough time, can do the work of any computer: the universal Turing machine.

Instead of asking whether or not a specific program halts, we look at the ensemble of all possible computer programs. We assign to each computer program a probability that it will be chosen. Each bit of information in the random program is chosen by tossing a coin, an independent toss for each bit, so that a program containing so many bits of information, say, N bits, will have a probability of 2^{-N}. We can now ask what is the total probability that those programs will halt. This halting probability, call it Omega, wraps up Turing's question of whether a program halts into one number between 0 and 1. If the program never halts, Omega is 0; if it always halts, Omega is 1.

In the same way that computers express numbers in binary notation, we can describe Omega in terms of a string of 1s and 0s. Can we determine whether the Nth bit in the string is a 0 or a 1? In other words, can we compute Omega? Not at all. In fact, I can show that the sequence of 0s and 1s is random. This is possible using what is called algorithmic information theory, which ascribes a degree of order in a set of information or data according to whether there is an algorithm that will compress the data into a briefer form.

For example, a regular string of 1s and 0s describing

some data such as 0101010101... which continues for 1,000 digits can be encapsulated in a shorter instruction 'repeat "01" 500 times'. A completely random string of digits cannot be reduced to a shorter program at all. It is said to be algorithmically incompressible.

My analysis shows that the halting probability is algorithmically random. It cannot be compressed into a shorter program. To get N bits of the number out of a computer, you need to put in a program at least N bits long. Each of the N bits of Omega is an irreducible independent mathematical fact, as random as tossing a coin. For example, there are as many 0s in Omega as 1s. And knowing all the even bits does not help us to know any of the odd bits.

My result that the halting probability is random corresponds to Turing's assertion that the halting problem is undecidable. It has turned out to provide a good way to give an example of randomness in number theory, the bedrock of mathematics.

The key to this was a dramatic development in the 1980s. James Jones of the University of Calgary in Canada and Yuri Matijasevic of the Steklov Institute of Mathematics in Leningrad discovered a theorem proved by Edouard Lucas in France a century ago. The theorem provides a particularly natural way to translate a universal Turing machine into a universal diophantine equation that is equivalent to a general purpose computer.

I thought it would be fun to write it down. So with the help of a large computer I wrote down a universal-Turing-machine equation. It had 17,000 variables and went on for 200 pages.

The equation is of a type that is referred to as

'exponential diophantine'. All the variables and constants in it are non-negative integers, 0, 1, 2, 3, 4, 5, and so on. It is called 'exponential' because it contains numbers raised to an integer power. In normal diophantine equations the power has to be a constant. In this equation, the power can be a variable. So in addition to having X^3, it also contains X^Y.

To convert the assertion that the halting probability Omega is random into an assertion about the randomness of solutions in arithmetic, I need only to make a few minor changes in this 200-page universal-Turing-machine diophantine equation. The result, my equation exhibiting randomness, is also 200 pages long. The equation has a single parameter, the variable N. For any particular value of this parameter, I ask the question: 'Does my equation have a finite or infinite number of whole-number solutions?' Answering this question turns out to be equivalent to calculating the halting probability. The answer 'encodes' in arithmetical language whether the Nth bit of Omega is a 0 or a 1. If the Nth bit of Omega is a 0, then my equation for that particular value of N has a finite number of solutions. If the Nth bit of the halting probability Omega is a 1, then this equation for that value of the parameter N has an infinite number of solutions. Just as the Nth bit of Omega is random – an independent, irreducible fact like tossing a coin – so is deciding whether the number of solutions of my equation is finite or infinite. We can never know.

To find out whether the number of solutions is finite or infinite in particular cases, say, for K values of the parameter N, we would have to postulate the K answers as K additional independent axioms. We would have to put in K bits of information into our system of axioms, so we would be no further forward. This is another way of saying that the K bits of information are irreducible mathematical facts.

I have found an extreme form of randomness, of irreducibility, in pure mathematics – in a part of elementary number theory that goes back 2,000 years to classical Greek mathematics. Hilbert believed that mathematical truth was black or white, that something was either true or false. I think that my work makes things look grey, and that mathematicians are joining the company of their theoretical physics colleagues. I do not think that this is necessarily bad. We have seen that in classical and quantum physics, randomness and unpredictability are fundamental. I believe that these concepts are also found at the very heart of pure mathematics.

4

My universe, my rules

A philosophical interlude

After all we have learned so far, let's take a moment or two to consider what it might mean. We like to think we're in control of ourselves, and slowly making sense of the universe around us. But what if neither of those things is possible? As we dig deeper into how much of a role chance plays in our lives, there is a possibility we will uncover some uncomfortable truths. Uncomfortable, but undeniably fascinating: your free will is in doubt, for instance. What's more, the laws of physics may ultimately prove unknowable. In fact, the very fundamental processes of the universe may be random chance – and for good reason: that may restore your free will, keep the future of the universe from being predetermined and stop information coming to you from the future. At the risk of inducing an existential crisis, it's time to spend a moment or two in philosophical indulgence.

Who's in charge here?

Do you ever wonder if your decisions are really yours? Getting to the bottom of why we do what we do is enough to keep anyone awake at night – or so you might think. But physicist Vlatko Vedral used to ponder this philosophical conundrum to get to sleep. Here, he explains why – and why, decades later, he is still wrestling with the puzzle of free will.

When I was a child I liked to ponder deep questions before falling asleep. One of my favourites was 'Do we have free will?' Shifting my mind back and forth between the possibilities served me well – it was a good technique for drifting off. Now I am a grown man I am lucky enough to have a job that involves deliberating over questions like this. So what does a man of science have to say about it?

Most of us in the west are certain that we have free will, though how we reach that conclusion, and even what we mean by it, is far from clear. If we define free will in everyday terms – as the capacity that allows us to control our actions – the answer would seem to come down to one of two possibilities: 'Yes, we do have free will', or 'No, we don't'. Both, however, quickly lead us into contradictions.

Suppose you answer with 'yes'. How would you demonstrate the validity of this statement? You would need to act in a way that would not be predetermined by anything. But how can this ever be, when whatever you do can be traced back to some cause?

Say you decide to prove your free will by acting out of character: having an introverted personality, you decide to start a conversation with a complete stranger on the street, for instance. The very fact that you have decided to act contrary to your usual predisposition seems itself to be fully predetermined by the fact that you wanted to act out of character to prove your free will. The very act of trying to prove free will adds to the evidence that you have none.

It appears that since we can never demonstrate free will we are forced to conclude that we do not have it. But this does not necessarily follow either – and feels completely contrary to the whole of human psychology. We reward people for doing good deeds and punish them for bad ones. This would seem to be completely misconceived if humans did indeed not have any free will. How can you punish someone for doing something when they are not free to do otherwise? Is our whole moral and judicial system based on an illusion?

This just cannot be the case – or at least it is impossible to live with. Can nothing good that I do be attributed to me? Is it all predetermined by my genes or my history or my parents or social order or the rest of the universe? It seems that we have no choice but to believe in free will.

We have now come full circle, from free will to no free will and back to free will again. Different strands of religion, which have also argued for both 'yes' and 'no' answers, don't help much. Eastern religions have come up with the notion of karma, according to which all your actions will lead to consequences in the interconnected universe. This has a deterministic ring to it, but then according to a number of eastern religions you can change your karma by acting differently. So it is up to you to change things, and you have free will to do so. In the west, in the Catholic version of Christianity, everyone is similarly predetermined by original sin, but can also use free will to do good deeds and end up in heaven: born to lose, but live to win. Most religions are about morality, and it would seem that assuming humans have free will is an essential part of that.

Science, on the other hand, seems to have its roots firmly set in the deterministic traditions. The main question for a physicist is: if I do so and so to such and such a system, how will it behave as a result of my actions? Physics is all about studying causes and their effects, so the deterministic link between cause and effect would seem to be of paramount importance. But is it?

Physics studies properties and laws of interaction of matter and energy. The ancient Greeks, especially philosophers such as Democritus and Leucippus, came up with the notion that everything in the world can be explained as collisions of the indivisible constituents of matter called atoms. This is identical in spirit, if not in form, to classical Newtonian physics and seems to leave no room for free will.

Newton imagined the universe to be like a giant clockwork machine running according to his immutable laws of motion. Not only is there no place for free will in Newton's universe, but even God plays a passive and marginal role. He is responsible only for setting the initial conditions. From then on the universe evolves deterministically without his intervention. Even when Newtonian gravity failed and Einstein replaced it with general relativity, nothing changed as far as determinism was concerned.

According to Einstein the universe actually exists all at once, and everything that has happened and will happen is already there in what we now call the 'block universe'. All the future instances of time are already laid out on a line in a four-dimensional block-like reality, as far as general relativity is concerned. Einstein is famously quoted as saying that any change with the passage of time is merely 'an illusion, albeit a persistent one'. This is full determinism at its best.

Quantum physics, however, has changed this picture in the most dramatic way. In quantum physics the notion of chance enters at a fundamental level. When a quantum particle, such as a particle of light – a photon – encounters a piece of glass such as your window, it seems to behave randomly. There is a chance that it will go through, but there is also a chance that it will be reflected. As far as we can tell there is nothing in the universe that determines which alternative will happen at any given time. The event of a photon being transmitted or reflected is, according to our best understanding of the laws of physics, a genuinely random event. The whole philosophy of one of the fathers of quantum theory, Niels Bohr, was founded on the assumption that the chance element is crucial to the nature of reality.

Einstein remained very much against this. Acknowledging randomness in quantum mechanics while having determinism in relativity would mean that these two pillars of physics can never be unified to describe the same reality. However, that's not the whole story. There is an interpretation of quantum mechanics according to which both determinism and randomness can be maintained in the quantum world. According to the 'many worlds' interpretation of quantum mechanics, all the alternatives, such as the photon being transmitted or reflected, exist at the same time, in the same universe – but in different worlds. So in one world the photon goes through your window, while in another 'parallel' world it is reflected.

In this picture we believe both worlds to be 'entangled' in their simultaneous existence within an overarching universe. And you, the observer of the photon, are also entangled within these worlds: there is a copy of you in one world who observes the photon going straight through the window, and another copy of you in another world observing the reflection of the photon from the surface of the window. Both versions of you, according to this interpretation, exist simultaneously in the same universe.

This is, of course, fully deterministic: everything that can happen does in fact happen. What you cannot decide, however – and here is where the element of chance is fundamental – is which particular world you yourself will occupy: which 'you' is you, and which 'you' is a copy. This kind of logic has led some people to conclude that consciousness may be fundamental for quantum mechanics, though it is not a conclusion I agree with.

In the end, however, it is clear that neither determinism nor randomness is good for free will. If nature is fundamentally random, then the outcomes of our actions are also completely beyond our control: randomness is just as bad as determinism.

To have the kind of free will we would like involves walking a fine line between determinism and randomness. We must be able to freely make our actions, but they should then result in deterministic (that is, non-random) effects. For example, we may want to be free to send our kids to a school of our choice. But then we also want to believe that the laws of physics (and biology, sociology and so on) ensure that going to a good school is highly likely to lead to a better life. Having free will is pointless without a certain degree of determinism.

The same can be said about studying physics. I want to believe that the choice regarding which aspect of nature I want to study – whether I want to measure the position or velocity of a particle, for example – lies with me. But what I also want is some degree of deterministic behaviour in nature that would then permit me to infer laws of physics from any measurement that I choose to make. In fact, the only means we have for deducing the basic equations of quantum mechanics means that they are fully deterministic, just like those of Newtonian mechanics.

There is nothing mysterious or controversial about this, but look what happens when we apply this to ourselves. If we are all made up of atoms, and if atoms behave deterministically, then we too must be fully determined. We simply must share the same fate as the rest of the universe. When we look inside our brains, all we find are interconnected neurons, whose behaviour in turn is governed by their underlying molecular structure, which in turn is fully governed by the strict laws of quantum mechanics. Taking the argument to extremes, the laws of quantum mechanics ultimately determine how I deduce the laws of quantum mechanics, which appears to be a fully circular argument and therefore logically difficult to sustain.

'What proof is there that brutes are other than a superior race of marionettes,' the biologist Thomas Henry

Huxley asked more than a century ago, 'which eat without pleasure, cry without pain, desire nothing, know nothing, and only simulate intelligence?' There is no proof as far as physics is concerned. Physics is simply unable to resolve the question of free will, although, if anything, it probably leans towards determinism.

The most honest position for a scientist on the question of free will is definitely agnostic: I simply do not know. What I do know is that when I was asked to write about free will as a physicist I found the idea so exciting that I had no choice but to agree to take it on.

Your uncertain future

We all believe the universe unfolds according to a set of laws. So why are so many things unpredictable? The answer to that question is surprisingly deep, as Paul Davies explains.

All science is founded on the assumption that the physical world is ordered. The most powerful expression of this order is found in the laws of physics. Nobody knows where these laws come from, nor why they apparently operate universally and unfailingly, but we see them at work all around us: in the rhythm of night and day, the pattern of planetary motions, the regular ticking of a clock.

The ordered dependability of nature is not, however, ubiquitous. The vagaries of the weather, the devastation of an earthquake, or the fall of a meteorite seem to be arbitrary and fortuitous. Small wonder our ancestors attributed these events to the moodiness of the gods. But how are we to reconcile these apparently random 'acts of God' with the supposed underlying lawfulness of the universe?

The ancient Greek philosophers regarded the world as a battleground between the forces of order, producing cosmos, and those of disorder, which led to chaos. They believed that random or disordering processes were negative, evil influences. Today, we don't regard the role of chance in nature as malicious, merely as blind. A chance event may act constructively, as in biological evolution, or destructively, such as when an aircraft fails from metal fatigue.

Though individual chance events may give the impression of lawlessness, disorderly processes, as a whole, may still display statistical regularities. Indeed, casino managers put as much faith in the laws of chance as engineers put in the laws of physics. But this raises something of a paradox. How can the same physical processes obey both the laws of physics and the laws of chance?

Following the formulation of the laws of mechanics by Isaac Newton in the 17th century, scientists became accustomed to thinking of the universe as a gigantic mechanism. The most extreme form of this doctrine was strikingly expounded by Pierre Simon de Laplace in the 19th century. He envisaged every particle of matter as unswervingly locked in the embrace of strict mathematical laws of motion. These laws dictated the behaviour of even the smallest atom in the most minute detail. Laplace argued that, given the state of the universe at any one instant, the entire cosmic future would be uniquely fixed, to infinite precision, by Newton's laws.

The concept of the universe as a strictly deterministic

machine governed by eternal laws profoundly influenced the scientific worldview, standing as it did in stark contrast to the old Aristotelian picture of the cosmos as a living organism. A machine can have no 'free will'; its future is rigidly determined from the beginning of time. Indeed, time ceases to have much physical significance in this picture, for the future is already contained in the present. The late Ilya Prigogine, a theoretical chemist at the University of Brussels and Nobel Laureate, eloquently expressed it thus: God is reduced to a mere archivist, turning the pages of a cosmic history book that is already written.

Implicit in this somewhat bleak mechanistic picture was the belief that there are actually no truly chance processes in nature. Events may appear to us to be random but, it was reasoned, this could be attributed to human ignorance about the details of the processes concerned. Take, for example, Brownian motion. A tiny particle suspended in a fluid can be observed to execute a haphazard zigzag movement as a result of the slightly uneven buffeting it suffers at the hands of the molecules of fluid that bombard it. Brownian motion is the archetypical random, unpredictable process. Yet, so the argument ran, if we could follow in detail the activities of all the individual molecules involved, Brownian motion would be every bit as predictable and deterministic as clockwork. The apparently random motion of the Brownian particle is attributed solely to our lack of information about the myriads of participating molecules, arising from the fact that our senses are too coarse to permit detailed observation at the molecular level.

For a while, it was commonly believed that apparently 'chance' events were always the result of our ignoring, or effectively averaging over, vast numbers of hidden variables, or degrees of freedom. The toss of a coin or dice, the spin of a roulette wheel – these would no longer appear random if we could observe the world at the molecular level. The slavish conformity of the cosmic machine ensured that lawfulness was folded up in even the most haphazard events, albeit in an awesomely convoluted tangle.

Two major developments of the 20th century have, however, put paid to the idea of a clockwork universe. First there was quantum mechanics. At the heart of quantum physics lies Heisenberg's uncertainty principle, which states that everything we can measure is subject to truly random fluctuations. Quantum fluctuations are not the result of human limitations or hidden degrees of freedom; they are inherent in the workings of nature on an atomic scale. For example, the exact moment of decay of a particular radioactive nucleus is intrinsically uncertain. An element of genuine unpredictability is thus injected into nature.

Despite the uncertainty principle, there remains a sense in which quantum mechanics is still a deterministic theory. Although the outcome of a particular quantum process might be undetermined, the relative probabilities of different outcomes evolve in a deterministic manner. What this means is that you cannot know in any particular case what will be the outcome of the 'throw of the quantum dice', but you can know with complete accuracy how the betting odds vary from moment to moment. As a statistical theory, quantum mechanics remains deterministic. Quantum physics thus builds chance into the very fabric of reality, but a vestige of the Newtonian–Laplacian worldview remains.

Then along came chaos theory. The essential ideas of chaos were already present in the work of the mathematician Henri Poincaré at the turn of the century, but it is only in recent years, especially with the advent of fast electronic computers, that people have appreciated the full significance of chaos theory.

The key feature of a chaotic process concerns the way that predictive errors evolve with time. Let me first give an example of a non-chaotic system: the motion of a simple pendulum. Imagine two identical pendulums swinging in exact synchronism. Suppose that one pendulum is slightly disturbed so that its motion gets a little out of step with the other pendulum. This discrepancy, or phase shift, remains small as the pendulums go on swinging.

Faced with the task of predicting the motion of a simple pendulum, one could measure the position and velocity of the bob at some instant, and use Newton's laws to compute the subsequent behaviour. Any error in the initial measurement propagates through the calculation and appears as an error in the prediction. For the simple pendulum, a small input error implies a small output error in the predictive computation. In a typical non-chaotic system, errors accumulate with time. Crucially, though, the errors grow only in proportion to the time (or perhaps a small power thereof), so they remain relatively manageable.

Now let me contrast this property with that of a

chaotic system. Here a small starting difference between two identical systems will rapidly grow. In fact, the hallmark of chaos is that the motions diverge exponentially fast. Translated into a prediction problem, this means that any input error multiplies itself at an escalating rate as a function of prediction time, so that before long it engulfs the calculation, and all predictive power is lost. Small input errors thus swell to calculation-wrecking size in very short order.

The distinction between chaotic and non-chaotic behaviour is well illustrated by the case of the spherical pendulum: a pendulum free to swing in two directions. In practice, this could be a ball suspended on the end of a string. If the system is driven in a plane by a periodic motion applied to the pivot, it will start to swing about. After a while, it may settle into a stable and entirely predictable pattern of motion, in which the bob traces out an elliptical path with the driving frequency. However, if you alter the driving frequency slightly, this regular motion may give way to chaos, with the bob swinging first this way and then that, doing a few clockwise turns, then a few anticlockwise turns in an apparently random manner.

The randomness of this system does not arise from the effect of myriads of hidden degrees of freedom. Indeed, by modelling mathematically only the three observed degrees of freedom (the three possible directions of motion), one may show that the behaviour of the pendulum is nonetheless random. And this is in spite of the fact that the mathematical model concerned is strictly deterministic.

It used to be supposed that determinism went hand

in hand with predictability, but we can now see that this need not be the case. A deterministic system is one in which future states are completely determined, through some dynamical law, by preceding states. But determinism implies predictability only in the idealised limit of infinite precision. In the case of the pendulum, for example, the behaviour will be determined uniquely by the initial conditions. The initial data includes the position of the bob, so exact predictability demands that we must assign a real number to the position that correctly describes the distance of the bob's centre from a fixed point. And this infinite precision is impossible.

Any predictive computation will contain some input errors because we cannot measure physical quantities to unlimited precision. Moreover, computers can handle only finite quantities of data.

In a non-chaotic system this limitation is not so serious because the errors expand only slowly. But in a chaotic system errors grow at an accelerating rate. Suppose there is an uncertainty in, say, the fifth significant figure, and that this affects the prediction of how the system is behaving after a time, t. A more accurate analysis might reduce the uncertainty to the tenth significant figure. But the exponential nature of error growth implies that the uncertainty now manifests itself after a time 2t. So a hundred-thousandfold improvement in initial accuracy achieves a mere doubling of the predictability span. It is this 'sensitivity to initial conditions' that leads to well-known statements about the flapping of butterflies' wings in the Amazon jungle causing a tornado in Texas.

Chaos evidently provides us with a bridge between

the laws of physics and the laws of chance. In a sense, chance or random events can indeed always be traced to ignorance about details – except, perhaps, in quantum theory. But whereas Brownian motion appears random because of the enormous number of degrees of freedom we are voluntarily overlooking, deterministic chaos appears random because we are necessarily ignorant of the ultra-fine detail of just a few degrees of freedom. And whereas Brownian chaos is complicated because the molecular bombardment is itself a complicated process, the motion of, say, the spherical pendulum is complicated even though the system itself is very simple. Thus, complicated behaviour does not necessarily imply complicated forces or laws. So the study of chaos has revealed how it is possible to reconcile the complexity of a physical world displaying haphazard and capricious behaviour with the order and simplicity of underlying laws of nature.

What can we conclude about Laplace's image of a clockwork world that contains a wide range of both chaotic and non-chaotic systems? Those that are chaotic have severely limited predictability, and even one such system would rapidly exhaust the entire universe's capacity to compute its behaviour. It seems, then, that the universe is incapable of digitally computing the future behaviour of even a small part of itself. Expressed more dramatically, the universe is its own fastest simulator.

This conclusion is surely profound. It means that, even accepting a strictly deterministic account of nature, the future states of the universe are in some sense 'open'. Some people have seized on this openness to argue for the reality of human free will. Others claim that it bestows upon nature an element of creativity, an ability to bring forth that which is genuinely new, something not already implicit in earlier states of the universe, save in the idealised fiction of the real numbers. Whatever the merits of such sweeping claims, it seems safe to conclude from the study of chaos that the future of the universe is not irredeemably fixed. To paraphrase Prigogine, the final chapter of the great cosmic book has yet to be written.

God plays dice – and for good reason

What is the point of a quantum-mechanical universe? Could it be any other way? Most of science is concerned with 'how' questions, but the ones asking 'why?' are in many ways even more interesting. Here, Mark Buchanan delves into the mind of God and finds that, when you're designing a universe, quantum is a good way to go.

This is a story about God – about His Intentions, and His Limitations. But it is not about religion. When Albert Einstein said 'God does not play dice' there was no thought of spiritual transcendence. He was simply expressing contempt for the notion that randomness might be an inherent part of whatever spirit, urge or process is behind our universe. But perhaps we can go even deeper.

Nearly a century after we first glimpsed the quantum nature of our world, the details of quantum events remain utterly unpredictable. So we may as well admit it: from atomic transitions to nuclear decays, the world really does seem to be random. God really does play dice. But why?

Why is the universe quantum mechanical? What's the point?

Physicists usually ask 'how' questions. How do photons and electrons pull off the quantum trick of being in many places at once? How does measuring them mysteriously cause them to make up their minds? By transforming 'how' into 'why', we can sidestep this impenetrable forest of quantum weirdness. And, although it is an audacious question, we are getting some clues and leads – and perhaps even the embryo of an answer. The randomness in the quantum world, some physicists now suspect, has a purpose. If they are right, the effect of quantum uncertainty is not to build chaos and disorder into our world. Quite the opposite. God uses it to ensure that all of the universe's far-flung regions remain a coherent part of His overall plan.

That paradoxical conclusion follows from studies of one of the weirdest of all quantum happenings, a phenomenon known as 'quantum entanglement'. Entanglement is an unnerving kind of link that can develop between two or more photons, electrons or atoms, even if they inhabit distant parts of the universe. Consider, for example, a pion, a subatomic particle which can decay into an electron and its antiparticle, a positron. When this happens, the particles fly off in opposite directions. But according to quantum theory, no matter how far apart the particles get, they remain mysteriously connected.

One of the oddities of quantum particles is that their properties only take on definite values when measured. The electron and positron, for instance, are both effectively spinning. The spin of either particle is equally likely to be clockwise (known as 'up') or anticlockwise ('down') – but you won't know which unless you measure it. Until that measurement is made the particle is in a weird indefinite state, a 'superposition' of both spins. What is definite, however, is that in an entangled state, the spins of the two particles are intimately linked. Since the original pion had no spin, the positron and electron must always spin in opposite senses so that their net spin remains zero. If you find the electron's spin to be 'up', you'll find the positron's to be 'down', and vice versa.

So it is as if the two entangled particles, no matter how far apart they are, are not really separate at all. Measure one, and as its spin becomes definite this triggers the other to respond. Its indeterminate spin also becomes definite, in the opposite direction to that of its partner. What is astonishing and disturbing is that this response happens instantaneously – even if the particles are separated by huge distances.

Consequently, quantum theory requires action at a distance. What happens in one part of the universe can have instantaneous 'non-local' consequences in other parts, no matter how far away they might be. And this poses a problem, because instantaneous action at a distance is a punch in the nose for Einstein. His theory of relativity – the cornerstone of physics – claims that our universe has an absolute speed limit. Nothing, according to Einstein, can travel faster than light.

So you might wonder – do we really need to swallow this non-local quantum weirdness? Perhaps there is a better theory that accounts for these entanglements without action at a distance?

Think of this: if someone separated a pair of your shoes by a great distance and then weighed one, they would immediately have a good estimate of the weight of the other. There's no mystery here. Nothing non-local. Shoes have weight. And if they come from a pair, their weights are correlated from the outset. Could something similar be true for entangled particle pairs? Despite what quantum theory says, perhaps the particles do have definite spins, arranged oppositely at all times, and measurements merely reflect this pre-existing situation.

This is an obvious possibility. It might even be true. The trouble is, it doesn't cushion the blow for relativity. We know that because, in 1964, physicist John Bell of CERN, the European Laboratory for Particle Physics, examined this line of argument in detail and proved a famous theorem which fellow physicist Henry Stapp, now retired from the Lawrence Berkeley Laboratory in California, calls 'the greatest discovery of all science'.

Bell first supposed that quantum theory doesn't say all there is to say about quantum particles. He then proved that if any more complete theory – any theory imaginable – were to give predictions in agreement with quantum theory, it would necessarily still contain the same kind of non-local influences as ordinary quantum theory. 'What Bell gave us,' says philosopher David Albert of Columbia University in New York, 'is a proof that there is a genuine non-locality in the workings of nature, however we attempt to describe it, period.' Every conceivable story about entangled states has to be non-local. There is no escape. Unless, of course, entangled states don't really exist, and quantum theory is wrong.

But we can be pretty sure it isn't wrong because there are experiments to prove it. In 1981, Alain Aspect of the Institute of Optics in Palaiseau, France, showed using pairs of photons that entanglement works just as quantum theory says it does. Other researchers have since improved on Aspect's results. Nicolas Gisin and his colleagues at the University of Geneva have used photon pairs that travelled inside fibre-optic cables to separate cities in Switzerland to show that entanglement can persist even for particles separated by tens of kilometres. Others have sent photons though the air over more than 100 kilometres to check these strange links. Distance is, it seems, irrelevant.

What's more, entanglement does not only apply to pairs of particles. University of Bristol mathematician Noah Linden, working with his colleague, theoretical physicist Sandu Popescu, has investigated entanglement between larger numbers of particles. They found that in the typical quantum state occupied by any group of particles the links between the particles are mostly of a non-local character. Quantum theory isn't just a tiny bit non-local. It's overwhelmingly non-local. Non-locality is the rule for our universe.

That is an unsettling conclusion. Non-locality cuts into the idea of the separateness of things, and threatens to ruin the very notion of isolation. To isolate an object we ordinarily move it a long way away from everything else, or build impenetrable walls around it. But the link of entanglement knows no boundaries. It isn't a cord running through space, but lives somehow outside space. It goes through walls, and pays no attention to distance.

Does this mean the idea of separateness is doomed?

And if faster-than-light connections are possible, is relativity – despite its huge successes – doomed too?

This is where God's dice-playing comes in. Popescu believes that the randomness at the heart of quantum mechanics is God's safeguard against such grotesque consequences. It ensures what physicist Abner Shimony of Boston University calls the 'peaceful coexistence' of quantum theory and relativity. Sure, the outcome, up or down, at one end of an entangled link instantaneously alters what happens at the other end. But the outcomes themselves are completely uncontrollable. No matter which particle you measure, you find the results up or down randomly, in equal measure. So you can't control the outcome at the other end. You can't use the link to send any kind of message.

And whatever tricks you try, this block on sending information instantaneously seems to remain unbreachable. Suppose you chose two separate axes, say A and B, on which to measure the spin of your particles. If you measured the spin of one particle on axis A, then its partner's spin on axis A would immediately be defined.

Likewise for spins on axis B. The fact that you can't control whether the spin is up or down would no longer matter. As long as you had some kind of device to tell you on which axis the spin had been defined you would have a way of sending a binary code: ABBABBAB, for example, would convey the same information as the conventional digital byte 01101101.

But it turns out that any conceivable detector capable of doing this is also prohibited by the mathematics of quantum theory. An experimenter at the other end can't possibly learn from individual outcomes, from the statistics of the outcomes, or from anything else, what was the sequence of your measurements. Quantum randomness prevents it.

So a stream of entangled particles is something like a combination of the most perfect telephone link and the most useless handsets you could ever imagine. The link can carry influences instantaneously across the universe. But the handsets at either end have the property that when you talk into them, they randomise your speech. 'Hello, it's me' you say, and into the line goes 'Nbsl Cvdibobo'. You can send a message faster than light, all right. You just can't extract any meaning from it when it arrives. Whatever goes from one particle to the other, as the late Asher Peres once put it, is 'information without information'.

According to Popescu, this answers the 'why' question. For despite the raw non-locality in the links of entanglement, randomness ensures that quantum theory doesn't transgress the letter of Einstein's law. At the core of Einstein's theory is the 'no-signalling' criterion: you cannot send energy or information from one place to another faster than light. This protects the chain of cause and effect, and ensures that effects never happen before their causes. In a deterministic world, any action at a distance would violate no-signalling. But quantum theory allows what Shimony calls 'passion-at-a-distance', a weaker linking up of distant things which stops just short of upsetting the principle of causality.

So the picture of God's world is this: through relativity, He ensures a degree of separateness and individuality for distant pieces of His Universe. Through quantum entanglement, He maintains links between distant regions, and keeps the whole universe coherently connected. It's the randomness that makes it possible for God to tie distant parts of the universe together more tightly than He otherwise could, while ensuring that cause and effect stay distinct. This is what He gets by playing dice.

'It is wondrous', says Popescu, 'that quantum mechanics combines non-locality and causality.' But he hopes to wring from his questions a bit more insight than that. In playing dice, does God have no other choice but to use quantum rules? 'Is quantum mechanics the only theory that can reconcile non-locality with relativity?' asks Popescu. If so, this might explain not only why the universe contains randomness, but why it enters the world in quantum-mechanical clothing.

In the early 1990s, Gisin explored the question of whether quantum theory could be modified in any way and still be consistent with what we know experimentally about the world. He found that fiddling with the edifice is an extraordinarily sensitive business. 'If you try to alter the theory very slightly, then quantum non-locality immediately becomes malignant: it can be used for faster-than-light signalling,' he says. But what if you alter it wildly? Is there any theory whatsoever besides the quantum theory in which non-locality and causality can coexist?

To find out, Popescu and his colleague Daniel Rohrlich of Ben Gurion University in Israel played some odd intellectual games. Their idea was to probe the realm of possible theories, and to consider alternative theories that go beyond quantum theory.

This isn't so much an exercise in physics as in

mathematics. It's not hard to dream up non-local theories. You can make up any number of them at will just by inventing forces that act at a distance between particles. However, most of these theories violate relativity by allowing faster-than-light signalling. It's also not hard to invent theories that respect no-signalling. Any theory with strictly local causes, for example, will do it. But the interesting theories are those that achieve both non-locality and no-signalling at once. Are there many theories like that? Or is quantum theory the only one?

Popescu and Rohrlich didn't have to go far to find an answer: quantum theory, they think, is not the only non-local theory with no-signalling. Their proof comes in the form of a model world they have constructed, in which particles can be entangled even more strongly than they are in the quantum world. This super-entanglement leads to 'supercorrelations' between spin measurements. And yet the physics still doesn't violate no-signalling. So this hypothetical world provides a proof of principle: there are other inhabitants of the weird theoretical terrain where non-locality and causality can coexist.

This doesn't mean that quantum theory is about to be ousted by one of these alternatives: in our world, quantum theory unquestionably rules. But the mere existence of these theories means that the need to have both non-locality and causality is not enough to tie God's hands and fix the laws of physics. There has to be something else.

'Our models raise a question,' says Popescu. 'What is the minimal set of principles – non-locality plus no-signalling plus something else, simple and fundamental, from which we could derive quantum mechanics?' Is there something we don't yet know about, some other principle as deep and pervasive as both causality and non-locality?

So while we may know why God plays dice, we don't yet know why he throws them as he does. Why quantum mechanical dice? What else constrains His hands? More bold questions for physicists and philosophers to get their teeth into.

5

Biology's casino

Chance in the natural world

Biological luck didn't just get us where we are today. It also shapes where the natural world will be tomorrow. Survival of the fittest can be determined by an ability to generate randomness – maybe to evade a predator, or to predict how a pathogen might evolve. Without the flexibility that random mutations provide, life might not survive Earth's future challenges. And as for you – the fact that you can pursue rational thoughts is because your brain is stimulated by random signals. The chances are, you couldn't think without chance.

A chance at life

Evolution often has adverse consequences for someone or something – better to make sure it's not us. Get to grips with chance mutations, and we can protect ourselves from life's twists; do it well, and we might even get a sneak preview of alien life, as Bob Holmes explains.

Take 100 newly formed planets of one Earth mass. Place each in the habitable zone of a G-type main sequence star. Set your timer for 4 billion years. What do you get? A hundred planets teeming with life forms quite similar to those on Earth, perhaps even dominated by naked apes? Or would evolution produce very different outcomes every time, if life even got started at all?

Some biologists argue that evolution is a deterministic process, that similar environments will tend to produce similar outcomes. Others, the most famous of whom was Stephen Jay Gould, think its course follows unpredictable twists and turns, and that the same starting point can lead to very different results.

The answer does matter. If the Gould camp is right, the study of evolution is like the study of history: something we can understand only in retrospect. If, however, the vagaries of chance play a minimal part, then biologists can predict the course of evolution to a large extent. That is an important difference because predicting evolution is crucial to stopping tumours becoming drug-resistant, or bacteria shrugging off an antibiotic, or bedbugs becoming immune to pesticides, or viruses killing people who have been vaccinated against them and so on.

So which is it? We might not have 100 Earths and a time machine, but we can look at how evolution has turned out on, say, neighbouring islands, or even rerun it over and over in the lab. These kinds of studies are giving us a better idea of the role of chance.

First things first. Evolution does begin with chance events, in the form of mutations. But it is not a case of anything goes; far from it. Which mutations survive and spread depends on natural selection – the survival of the fittest. Put another way, chance is the creative partner that comes up with all the ideas – some brilliant, others hopeless – while natural selection is the ruthlessly practical one, picking what works.

Many biologists, most notably Richard Dawkins, therefore insist that although mutations may be random, evolution is not. This insistence might make sense when explaining evolution to people who have not grasped the basic concept. But there is an element of chance in evolution, even when natural selection is firmly in the driving seat.

Take the evolution of flu viruses. We can predict with confidence that, over the next few years, the structure of a viral surface protein called haemagglutinin will evolve so that the human immune system can no longer recognise and attack it. What's more, we can even be fairly sure that the mutations that allow new strains of flu to evade the immune system will happen at one of seven critical sites in the gene coding for haemagglutinin, according to Trevor Bedford, an evolutionary biologist at Fred Hutchinson Cancer Research Center in Seattle. In this sense, the evolution of flu is non-random and predictable.

But it's a matter of chance which of those seven sites mutate, and how. Predicting the course of flu's evolution is almost impossible more than a year or two in advance, says Bedford. This is why flu vaccine makers do not always get it right, and why flu vaccines are sometimes largely ineffective.

What's more, important as natural selection is, its powers are limited. The fittest do not always survive;

instead, the course of evolution is often shaped by accidental events. If it hadn't been for an asteroid strike, for instance, we mammals might still be scurrying about in mortal fear of dinosaurs (see 'Asteroids with a silver lining', page 172). And if a different bird had been blown to the far-off Galapagos Islands a few million years ago, we might be talking about Darwin's crows instead of Darwin's finches.

We've long known that new populations can have low genetic variability. But studies suggest that this 'founder effect' may be more important than thought. For example, a handful of little birds were the ancestors of several populations of Berthelot's pipit on the Selvagem and Madeira island chains, in the Atlantic. There are big variations among them in the shape and size of beaks, legs and wings.

When Lewis Spurgin of the University of East Anglia in Norwich, UK studied these populations he expected to find environmental differences that explained this variation, but that is not what happened. Instead, he concluded that the physical differences were not driven by natural selection but were just a result of the small number of founders: accidents of history, in other words.

The founder effect can even create new species without the need for natural selection. When Daniel Matute, now at the University of North Carolina in Chapel Hill, took a large population of fruit flies and created 1,000 founder populations of a single male and female in identical vials in his lab, most populations simply became extinct because of inbreeding. But in three of the surviving populations, the founders produced offspring sufficiently different that they were less able to interbreed with the larger parental population – the first step to the creation of a new species.

Effects like these might explain why the islands of Hawaii have such a rich diversity of fruit flies. In fact, a few biologists think speciation is almost always an accidental process, rather than one driven by natural selection.

Yet more evidence of the limits of natural selection comes from genomes, which are littered with the products of chance. Despite many claims to the contrary, most of the human genome is just junk. This junk has accumulated because natural selection has not been strong enough to remove it, says Michael Lynch, an evolutionary biologist at Indiana University in Bloomington. In small populations, even mutations that are slightly harmful can spread throughout the population simply by chance.

Does this kind of genetic drift really matter? At least sometimes, it does. Joe Thornton of the University of Chicago has been turning back the clock and replaying evolution to see if it could have turned out differently. Think *Jurassic Park*, except rather than recreate extinct animals, Thornton has recreated ancient proteins. His team began with living vertebrates that each have their own version of the gene encoding the protein that detects the stress hormone cortisol. By comparing the versions, the team could work out how this protein had evolved over hundreds of millions of years, from a protein that could detect another hormone.

Then Thornton's team went a lot further. They actually made some of these ancient proteins and tried them out to see what effect each mutation had. Switching to cortisol took five mutations: two to recognise cortisol and three to 'forget' the previous hormone.

But when the team made only these five changes, they destabilised the protein and wrecked it. It turns out the transition to cortisol was possible only because two other mutations that stabilise the protein had occurred first. But these 'permissive' mutations have no effect by themselves. They must have arisen by chance, not by natural selection.

'We think of these permissive mutations as opening doors, so that evolution has the opportunity to follow pathways that were inaccessible without the permissive mutations,' says Thornton. And there seems to be only one way the door to the cortisol-binding pathway could have opened. Thornton tested thousands of other mutations, but none did the trick. 'There is nothing else in the neighbourhood around the ancestral protein that could have opened that door,' he says.

In Thornton's view, the course of evolution often – although not always – hinges on such seemingly insignificant chance events. In this way, evolution is a lot like life, he notes: a seemingly inconsequential decision one night to go to one party rather than another might lead to meeting your future partner and thus change the course of your life.

Then again, who we hook up with seldom alters the course of history. Although all these studies suggest that chance plays a bigger role in evolution than generally acknowledged, the big question is how much difference it makes in the long run. The detailed paths taken by evolving populations might depend largely on chance, yet still lead to similar outcomes. There are only so many ways of flying and swimming, for instance, which is why wings and fins have independently evolved on many occasions. If Thornton's protein hadn't evolved the ability to bind cortisol, perhaps another protein would have instead.

There are many examples of this kind of convergent evolution. Arctic and Antarctic fish have independently evolved antifreeze proteins that work in the same way, for example, while several snake lineages have separately come up with identical methods of coping with the poisons secreted by the newts they eat.

In the Greater Antilles in the Caribbean, meanwhile, evolution has effectively been rerun on four islands – and turned out the same way. Each of the islands has long-legged Anolis lizards that run on the ground, short-legged ones that grasp twigs, and lizards with big toepads that stick to leaves. But each island's lizards seem to derive from a single founder population, meaning they independently evolved to fill the same niches.

Does this mean Gould was wrong after all, that in the long run chance does not matter that much? Perhaps the closest we can get to an answer is the Long-Term Experimental Evolution Project, led by Richard Lenski of Michigan State University. On 24 February 1988, Lenski took samples of one kind of E. coli bacteria and used them to found 12 new populations. Every day since then – on weekends and holidays, despite blizzards and grant deadlines – someone has kept them growing by transferring samples to a new nutrient medium.

In the almost three decades that have passed, Lenski's populations have evolved for over 62,000 generations.

For comparison, *Homo sapiens* has gone through perhaps 20,000 generations in its entire existence. All 12 populations have changed in similar ways, evolving larger cells and faster growth rates, showing that sometimes evolution really does unfold in predictable ways.

But even without external events like asteroid strikes, the course of Lenski's populations was not always predictable. One population evolved into a mix of two lineages, each of which survives because it pursues slightly different strategies. Another suddenly developed, at about the 31,500th generation, the ability to feed on citrate, an additive to the culture medium that *E. coli* cannot normally use. 'They started from the same place and were subjected to exactly the same conditions, and differences still pop up,' says Zachary Blount, who works with Lenski on the project. 'The differences arise purely out of the chance that is inherent in the evolutionary process.'

Was the citrate-using mutation a lucky break, or could evolution find it again? Because Lenski's team freezes a sample of each culture every 500 generations, Blount was able to go back into the archives of this population and literally rerun evolution. When he did so, the only time citrate use evolved was when he began the replay with cells from the 20,000th generation or later.

Clearly, some mutation or mutations must have happened around the 20,000th generation that set the stage for citrate use to evolve much later, just as Thornton's hormone receptor required permissive mutations before it could switch to recognise a different target. 'We still haven't figured out what that mutation was, which is really frustrating,' says Blount. Until they can, the team will not know whether the permissive mutation offered some other advantage to the bacteria. Even if it did, however, it seems clear that its role in permitting citrate use must have been just a lucky by-product.

So what would we get if we could replay evolution over and over on a planetary scale? One possibility is an awful lot of slime. Nick Lane of University College London thinks that the emergence of complex cells depended on a highly unlikely merger of two kinds of simple cell. If he's right, bacteria-like life forms could be common on other worlds but rarely give rise to more sophisticated organisms.

But assuming life did get past the slime stage on other worlds, what would it be like? 'There is a fairly good chance that such replays would often yield worlds that look broadly like ours in terms of what niches are filled, and what sorts of major traits you see,' says Blount. In other words, you'd still expect to see photosynthesisers and predators, and parasites and decomposers. But the details are likely to differ sharply from one replay to the next, according to Blount. Even if we replayed evolution a hundred times, it's highly unlikely that we would end up again with a big-brained primate ruling the planet.

But would some other brainy, social animal take over the planet? Maybe. 'There's clearly an adaptive zone in most habitats that involves intelligence,' says David Jablonski, a palaeontologist at the University of Chicago. And it has become clear that many traits we once thought of as uniquely human, from language to tool-making, exist to some extent in many other animals. So although naked apes might not emerge on any of the 100 planets, other smart tool-users might.

It is a question we might even be able to answer one day. Thousands of exoplanets have now been discovered and even though we've yet to find any just like ours, all the evidence suggests there are plenty of Earth-like planets close enough that we might not only determine whether they support life, but also learn a little about it. The answer may be in the stars.

Asteroids with a silver lining

Every 100 million years or so something big wallops the planet. If it happened now we would be wiped out. Yet curiously, we probably owe our existence to the last such impact.

Around 65.5 million years ago an asteroid some 10 kilometres across slammed into the Yucatan peninsula in present-day Mexico. The release of carbon and sulphur-rich gases from the blasted rock layers precipitated a global catastrophe in which fires raged, the sky darkened, Earth cooled and acid rain showered down. Within months the dinosaurs were dead. So too were almost all other reptiles of sea and air, along with ammonites and most birds and land plants.

For mammals, it was a different story. They didn't exactly sail through – around half of the species went extinct – but those that survived were small, fast-breeding and versatile, and could scavenge the abundant detritus created by the impact. They were able to burrow or hide to escape the fires and acid rain. They often lived in or around freshwater ecosystems, which are fed largely by dead organic matter and so were more resilient in the face of catastrophe than oceans and dry land.

These survivors went on to inherit the Earth. As the biosphere gradually recovered, mammals filled the niches left vacant by the dinosaurs, and eventually those of the marine reptiles too. The fossil record suggests this happened in a burst of evolutionary creativity between 65 and 55 million years ago. Some 'molecular clock' studies, which compare the genomes of related living species to reconstruct their evolutionary tree, paint a slightly different picture, implying that mammalian evolution didn't gear up until more than 10 million years after the impact.

Either way, one lineage that makes its debut is ours, the primates. Reason enough to say that if that asteroid had not been there and then, we would not be here and now

Graham Lawton

Bullet-proof

Is life using a built-in randomness generator that allows it to survive any scenario? It's a controversial idea because it invokes 'epigenetic' ideas of traits inherited without being encoded in DNA. Nonetheless, we know that evolution will use every available trick – including plenty of once-controversial ideas that are now firmly in the mainstream. Henry Nicholls takes up the story.

A man walks into a bar. 'I have a new way of looking at evolution,' he announces. 'Do you have something I could write it down on?' The barman produces a piece of paper and a pen without so much as a smile. But then, the man wasn't joking.

The man in question is Andrew Feinberg, a leading geneticist at Johns Hopkins University in Baltimore; the bar is The Hung, Drawn and Quartered, a pub within the shadow of the Tower of London. What's written on the piece of paper could fundamentally alter the way we think about epigenetics, evolution and common diseases.

Before setting foot in the pub, Feinberg had taken a spin on the London Eye, climbed Big Ben and wandered into Westminster Abbey. There, as you might expect, he sought out the resting place of Isaac Newton and Charles Darwin. He was struck by the contrast between the lavish marble sculpture of a youthful Newton, reclining regally beneath a gold-leafed globe, and Darwin's minimalist floor stone.

As he looked round, Feinberg's eyes came to rest on a nearby plaque commemorating physicist Paul Dirac. This set him thinking about quantum theory and evolution, which led him to the idea that epigenetic changes – heritable changes that don't involve modifications to DNA sequences – might inject a Heisenberg-like uncertainty into the expression of genes, which would boost the chances of species surviving. That, more or less, is what he wrote on the piece of paper.

Put simply, Feinberg's idea is that life has a kind of built-in randomness generator which allows it to hedge its bets. For example, a characteristic such as piling on the fat could be very successful when famine is frequent, but a drawback in times of plenty. If the good times last for many generations, however, natural selection could eliminate the gene variant for piling on fat from a population. Then, when famine does eventually come, the population could be wiped out.

But if there is some uncertainty about the effect of

genes, some individuals might still pile on the fat, even though they have the same genes as everyone else. Such individuals might die young in good times, but if famine strikes they might be the only ones to survive. In an uncertain world, uncertainty could be crucial for the long-term survival of populations.

The implications of this idea are profound. We already know there is a genetic lottery – every fertilised human egg contains hundreds of new mutations. Most of these have no effect whatsoever, but a few can be beneficial or harmful. If Feinberg is right, there is also an epigenetic lottery: some people are more (or less) likely to develop cancer, drop dead of a heart attack or suffer from mental health problems than others with exactly the same DNA.

To grasp the significance of Feinberg's idea, we have to briefly rewind to the early 19th century, when the French zoologist Jean-Baptiste Lamarck articulated the idea – already commonly held – that 'acquired characteristics' can be passed from parent to offspring. If a giraffe kept trying to stretch to reach leaves, he believed, its neck would get longer, and its offspring would inherit longer necks.

Contrary to what many texts claim, Darwin believed something similar, that the conditions an organism experiences can lead to modifications that are inherited. According to Darwin's hypothesis of pangenesis, these acquired changes could be harmful as well as beneficial – such as sons getting gout because their fathers drank too much. Natural selection would favour the beneficial and weed out the harmful. In fact, Darwin believed acquired changes provided the variation essential for evolution by natural selection.

Pangenesis was never accepted, not even during Darwin's lifetime. In the 20th century it became clear that DNA is the basis of inheritance, and that mutations that alter DNA sequences are the source of the variation on which natural selection acts. Environmental factors such as radiation can cause mutations that are passed down to offspring, but their effect is random. Biologists rejected the idea that adaptations acquired during the life of an organism can be passed down.

Even during the last century, though, examples kept cropping up of traits passed down in a way that did not fit with the idea that inheritance was all about DNA. When pregnant rats are injected with the fungicide vinclozolin, for instance, the fertility of their male descendants is lowered for at least two generations, even though the fungicide does not alter the males' DNA.

No one now doubts that environmental factors can produce changes in the offspring of animals even when there is no change in DNA. Many different epigenetic mechanisms have been discovered, from the addition of temporary 'tags' to DNA or the proteins around which DNA is wrapped, to the presence of certain molecules in sperm or eggs.

What provokes fierce argument is the role that epigenetic changes play in evolution. A few biologists, most prominently Eva Jablonka of Tel Aviv University in Israel, think that inherited epigenetic changes triggered by the environment are adaptations. They describe these changes as 'neo-Lamarckian' and some even claim that such processes necessitate a major rethink of evolutionary theory.

While such views have received a lot of attention, most

biologists are far from convinced. They say the trouble with the idea that adaptive changes in parents can be passed down to offspring via epigenetic mechanisms is that, like genetic mutations, most inherited epigenetic changes acquired as a result of environmental factors have random and often harmful effects.

At most, the inheritance of acquired changes could be seen as a source of variation that is then acted on by natural selection – a view much closer to Darwin's idea of pangenesis than Lamarck's claim that the intent of an animal could shape the bodies of its offspring. But even this idea is problematic, because it is very rare for acquired changes to last longer than a generation.

While epigenetic changes can be passed down from cell to cell during the lifetime of an organism, they do not normally get passed down to the next generation. 'The process of producing germ cells usually wipes out epigenetic marks,' says Feinberg. 'You get a clean slate epigenetically.' And if epigenetic marks do not usually last long, it's hard to see how they can have a significant role in evolution – unless it is not their stability but their instability that counts.

Rather than being another way to code for specific characteristics, as biologists like Jablonka believe, Feinberg's 'new way of looking at evolution' sees epigenetic marks as introducing a degree of randomness into patterns of gene expression. In fluctuating environments, he suggests, lineages able to generate offspring with variable patterns of gene expression are most likely to last the evolutionary course.

Is this 'uncertainty hypothesis' right? There is evidence

that epigenetic changes, as opposed to genetic mutations or environmental factors, are responsible for a lot of variation in the characteristics of organisms. The marbled crayfish, for instance, shows a surprising variation in coloration, growth, lifespan, behaviour and other traits even when genetically identical animals are reared in identical conditions. And a 2010 study found substantial epigenetic differences between genetically identical human twins. On the basis of their findings, the researchers speculated that random epigenetic variations are actually 'much more important' than environmental factors when it comes to explaining the differences between twins.

More evidence comes from the work of Feinberg and his collaborator Rafael Irizarry, a biostatistician at Harvard University. One of the main epigenetic mechanisms is the addition of methyl groups (with the chemical formula CH_3) to DNA, and Feinberg and Irizarry have been studying patterns of DNA methylation in mice. 'The mice were from the same parents, from the same litter, eating the same food and water and living in the same cage,' Feinberg says.

Despite this, he and Irizarry were able to identify hundreds of sites across the genome where the methylation patterns within a given tissue differed hugely from one individual to the next. Interestingly, these variable regions appear to be present in humans too. 'Methylation can vary across individuals, across cell types, across cells within the same cell type and across time within the same cell,' says Irizarry.

It fell to Irizarry to produce a list of genes associated with each region that could, in theory at least, be affected by the variation in methylation. What he found blew him away. The genes that show a high degree of epigenetic plasticity are very much those that regulate basic development and body-plan formation. 'It's a counter-intuitive and stunning thing because you would not expect there to be that kind of variation in these very important patterning genes,' says Feinberg.

The results back the idea that epigenetic changes to DNA might blur the relationship between genotype (an organism's genetic make-up) and phenotype (its form and behaviour). 'It could help explain why there is so much variation in gene expression during development,' says Günter Wagner, an evolutionary biologist at Yale University. But that does not necessarily mean epigenetic changes are adaptive, he says. 'There has not been enough work on specifying the conditions under which this kind of mechanism might evolve.'

When he began exploring the idea with Feinberg, Irizarry constructed a computer simulation to help him get his head round it. At first, he modelled what would happen in a fixed environment where being tall is an advantage. 'The taller people survive more often, have more children and eventually everyone's tall,' he says.

Then, he modelled what would happen in a changeable environment where, at different times, it is advantageous to be tall or short. 'If you are a tall person that only has tall kids, then your family is going to go extinct.' In the long run, the only winners in this kind of scenario are those that produce offspring of variable height.

This result is not controversial. 'We know from theory that goes some way back that mechanisms that induce "random" phenotypic variation may be selected over those that produce a single phenotype,' says Tobias Uller, a developmental biologist at the University of Oxford. But showing that something is theoretically plausible is a long way from showing that the variability in methylation evolved because it boosts survival.

Jerry Coyne, an evolutionary geneticist at the University of Chicago, is blunter. 'There is not a shred of evidence that variation in methylation is adaptive, either within or between species,' he says. 'I know epigenetics is an interesting phenomenon, but it has been extended willy-nilly to evolution. We're nowhere near getting to grips with what epigenetics is all about. This might be a part of it, but if it is it's going to be a small part.'

To Susan Lindquist of the Massachusetts Institute of Technology, however, it is an exciting idea that makes perfect sense. 'It's not just that epigenetics influences traits, but that epigenetics creates greater variance in the traits and that creates greater phenotypic diversity,' she says. And greater phenotypic diversity means that a population has a better chance of surviving whatever life throws at it.

Lindquist studies prions, proteins that can not only flip between two states but pass on their state to other prions. While they are best known for causing diseases such as Creutzfeldt-Jakob disease, Lindquist thinks they provide another epigenetic mechanism for evolutionary 'bet-hedging'. Take Sup35, a protein involved in the protein-making machinery of cells. In yeast, Sup35 has a tendency to flip into a state in which it clumps together, spontaneously or in response to environmental stress, which in turn can alter the proteins that cells make, Lindquist says. Some of these changes will be harmful, but she and her colleagues have shown that they can allow yeast cells to survive conditions that would normally mean death.

While Jablonka remains convinced that epigenetic marks play an important role in evolution through 'neo-Lamarckian' inheritance, she welcomes Feinberg and Irizarry's work. 'It would be worth homing in on species that live in highly changeable environments,' she suggests. 'You would expect more methylation, more variability, and inheritance of variability from one generation to the next.'

As surprising as Feinberg's idea is, it does not challenge the mainstream view of evolution. 'It's straight population genetics,' says Coyne. Favourable mutations will still win out, even if there is a bit of fuzziness in their expression. And if Feinberg is right, what evolution has selected for are not epigenetic traits, but a genetically encoded mechanism for producing epigenetic variation. This might produce variation completely randomly or in response to environmental factors, or both.

Feinberg predicts that if the epigenetic variation produced by this mechanism is involved in disease, it will be most likely found in conditions like obesity and diabetes, where lineages with a mechanism for surviving environmental fluctuation would win out in the evolutionary long run. A few years ago, he, Irizarry and other colleagues studied DNA methylation in white blood cells collected in 1991 and 2002 from the same individuals in Iceland. From this, they were able to identify more than 200 variably methylated regions.

To see if these variable regions have something to do with human disease, they looked for a link between methylation density and body mass index. There was a correlation at four of these sites, each of them located either within or near genes known to regulate body mass or diabetes. Feinberg sees this in a positive light. If random epigenetic variation does play a significant role in determining people's risk of getting common diseases, he says, untangling the causes may be simpler than we thought. The key is to combine genetic analyses with epigenetic measurements.

Feinberg is the first to admit that his idea could be wrong. But he's excited enough to put it to the test. Perhaps, he suggests, it is the missing link in understanding the relationship between evolution, development and common disease. 'It could turn out to be really quite important,' he says.

Do dither

We tend to think of the random signals of noise as a problem, but in many systems – both biological and technological – noise is actually an opportunity. Laura Spinney reckons your brain might not function half as well without it.

Noise is usually a nuisance, as anyone who lives under a flight path or has tried to listen to a distant AM radio station can testify. But to engineers its random fluctuations can be a godsend.

During the Second World War, aircrews who had to

calculate mission routes and bomb trajectories found that their instruments performed better in the air than on the ground. Air force engineers soon worked out why. As the planes moved through the air, their airframes were vibrating at a wide range of frequencies. By chance, some of these frequencies matched the resonant frequencies of the various moving parts of the instruments and gave them a nudge, allowing them to move more freely. Not knowing which frequencies were important, the engineers began building small vibrating motors into the instruments in the hope of kicking off a resonance. This was one of the earliest applications of dither, or the deliberate addition of random noise.

Now we are discovering that evolution beat us to it: biology has already harnessed the benefits of random signals. Under some circumstances, a small injection of noise can sharpen up the way in which an organism senses its environment. Crayfish, for example, are better at detecting the subtle fin movements of predatory fish when the water is turbulent rather than still. Humans, it turns out, are better able to recognise a faint image on a screen when a dash of noise is added to it.

In these cases the noise source is external to the organism, but such cases raise an intriguing possibility: could evolution have incorporated dither into the brain itself? A group of neuroscientists is now claiming to have found just that, in the form of neural circuits that are 'noisy by design'. If this group is right, it may be that dither is a common feature in nature.

A working definition of noise is that it is a broadband signal containing a jumble of frequencies – the hiss of white noise, for example, is made up of the full range of audible frequencies, from very low to very high, in equal amounts. In contrast, meaningful signals concentrate their energy on a comparatively narrow band of the spectrum.

The phenomenon of noise increasing the detectability of a faint signal is called stochastic resonance. ('Stochastic' simply means randomly patterned.) Stochastic resonance applies specifically to non-linear systems, where the output is not proportional to the input. Neurons are a good example of a non-linear system, firing only when the electrical potential across their membrane reaches a critical threshold. In such a system, a weak input which fails to reach the threshold can be lifted above it by the injection of noise.

Numerous theoretical models suggest that stochastic resonance could improve how neurons process signals, and there is good experimental evidence that adding external noise can enhance the brain's abilities under certain circumstances. Stochastic resonance explains why water turbulence helps a crayfish's sensory hair cells detect a distant fin movement, and why noise helps the human eye to pick out a faint image. External noise has since been harnessed to enhance human performance, for example, in cochlear implants to help pick up faint sounds and in vibrating insoles that reduce swaying in people who have suffered strokes.

For a long time, however, no one found any evidence that the brain generates its own internal noise to exploit stochastic resonance. Then along came Gero Miesenböck, a neuroscientist at the University of Oxford. Miesenböck thinks he has found a brain circuit, part of the olfactory system of the fruit fly Drosophila, that exists specifically to generate noise and thus enhance brain function. He says his discovery has implications for the human brain because the basic architecture of the Drosophila olfactory system is common not only to all insects but also to all vertebrates – including humans.

Miesenböck didn't set out in search of noise. He was trying to solve a mystery that has troubled olfactory-system researchers for many years.

The fly olfactory system is a huge piece of neural circuitry. It starts in the fly's antennae with around 1,200 olfactory receptor neurons (ORNs), each of which carries a single type of odour-receptor molecule. There are about 60 different receptor molecules and hence about 60 different types of ORN.

From the antennae, these odour-specific ORNs converge on nodes called glomeruli where they make synaptic connections with cells called projection neurons. Each glomerulus receives inputs from only one type of ORN, so for a long time neuroscientists assumed that each projection neuron would only respond to a single odour.

But a few years ago, neuroscientists discovered that this is not the case. Electrical recordings from individual projection neurons show that they sometimes respond to odours other than those picked up by their ORNs.

But how do they do this, when each glomerulus receives inputs from only one type of ORN? While at Yale School of Medicine a few years ago, Miesenböck and his colleague Yuhua Shang managed to solve this puzzle.

They took a mutant fly in which all the ORNs connected to a particular glomerulus were missing, and looked for other inputs to the projection neuron. What they found was a previously unknown network of 'interneurons' connecting the glomeruli to each other and transmitting activity between them. These 'excitatory local neurons' seem to provide a sort of diffuse, stimulatory input to the projection neurons whenever an odour is present.

That solved the immediate problem, but raised another: why add something to the system that means losing the exquisite one-to-one mapping of odour receptors to projection neurons? 'It seems counter-intuitive,' says Miesenböck. 'Why would you take the crisp, sharply separated input and blur it out, make it noisier?' The hypothesis he came up with was that the noise was there for a reason. Perhaps the excitatory local neurons deliberately inject noise into the system, taking advantage of stochastic resonance to make faint odours easier to detect.

This makes sense in the light of what subsequently happens to the sensory input signal. Projection neurons send signals to other neurons called Kenyon cells in a structure called the mushroom body, a part of the fly's brain involved in learning and memory. Each Kenyon cell receives inputs from many projection neurons, but they have extremely high firing thresholds and are only activated when a large number of their incoming neurons fire simultaneously. Given that projection cells fire more readily in response to their own odour than others, each Kenyon cell only fires in response to a single odour and the system recaptures specificity.

Miesenböck's group also came across a 1983 paper by Alexander Borst of the Max Planck Institute of

Neurobiology in Martinsried, Germany, describing a network of inhibitory local neurons linking the glomeruli. Miesenböck thinks these may have the opposite effect to his excitatory ones, damping down strong signals from ORNs.

So why bother to boost weak signals and tone down strong ones? Miesenböck suggests this happens to iron out extremes in odour concentrations. 'You need to be able to smell a rose, and identify it as a rose, at very faint concentrations and in full bloom, if it is held directly under your nose,' he says. 'There has to be some mechanism that eliminates the variation based on odour concentration. We think that the middle layer of processing does exactly that.'

Miesenböck's group still has some way to go to prove the 'noisy by design' hypothesis, but they're working on it. By tinkering with local neurons, they hope to learn how to change the volume of the noise. Miesenböck predicts that turning it down or silencing it entirely will make faint odours less likely to trigger Kenyon cells. Another prediction is that the flies will become behaviourally less responsive to faint smells, which the researchers can test by looking at their avoidance of bad ones.

Manipulations of this kind are tricky, however, partly because the researchers have no idea how many local neurons there are in a Drosophila brain. They need to modify the majority of them if they are to see the effects they are looking for.

If they succeed they will then attempt to show that something similar is happening in the mammalian brain. But finding a noise-generating cell resembling a fly's local neuron in a mammalian brain will be a huge challenge, according to Thomas Klausberger, also at Oxford. Klausberger has discovered new kinds of interneuron in the rat hippocampus, a structure that has been compared to the insect mushroom body because of its role in learning and memory. He points out that one region alone contains at least 21 different types of interneuron.

The 1993 crayfish study was carried out by Frank Moss, a biophysicist working at the University of Missouri in St Louis. Moss long suspected that animals take advantage of stochastic resonance to boost their reproductive success and was impressed by Miesenböck's findings.

One of Moss's studies was the first to show that externally applied noise worked via stochastic resonance. He was working on paddlefish, which find food by using electrosensors in their snouts to detect faint electrical signals given off by plankton, their natural prey. Moss put a paddlefish in a tank of water containing plankton, along with two electrodes which generated noise in the form of a randomly varying electric field. When he measured the effect of the noise, he found that there was an intermediate amplitude at which the fish's success rate significantly increased.

Optimal performance when the noise level is intermediate is one of the characteristics of stochastic resonance: too little noise and the signal doesn't reach the threshold, too much and the signal will be swamped by noise. The noise–benefit relationship is therefore shaped like an inverted U.

Moss subsequently turned his attention to tiny aquatic crustaceans called Daphnia or water fleas. He believed that they provided another strand of evidence pointing to internally generated stochastic resonance.

Daphnia have a characteristic foraging behaviour that follows the sequence of a hop, a pause, a turn through an angle and another hop. The turn angles vary and appear random to the naked eye.

Moss thought otherwise. He and his colleagues videoed five different species of Daphnia as they foraged for food in a shallow tank, and measured hundreds of turning angles. When they plotted the frequency distribution of these angles, they found that it was not completely random: some turning angles were more frequent than others. Their overall distribution could be described mathematically using a parameter called 'noise intensity' – a measure of how random, or noisy, it is.

Next they ran computer simulations of foraging Daphnia using different noise intensities. They found that the most successful food-gathering strategy used the noise-intensity level they had measured in real Daphnia. Lower or higher noise intensities reduced foraging success according to the classic inverted-U shape of stochastic resonance. Though no one yet knows how Daphnia generate their distribution of turning angles, Moss's team argued that it is an example of stochastic resonance in action and that it must be produced somewhere within the Daphnia, maybe in its brain. He believed that the optimal noise intensity must be the product of natural selection, because Daphnia using it would find more food and so maximise their fitness.

The idea that biological systems exploit internally generated noise still has questions hanging over it, however.

One big one is whether what is being generated by the local neurons in the fruit fly is genuine noise. Bart Kosko, an electrical engineer at the University of Southern California, Los Angeles, and author of the 2006 book *Noise*, says he is not convinced it is.

Noise has a strict mathematical definition and what looks like noise in a complex biological system usually turns out to be a signal leaking from elsewhere. 'What needs to be done is to take that "noise" source and show that it has the statistical footprint of noise,' says Kosko. If it isn't genuine noise then, by definition, you haven't got stochastic resonance.

Neuroscientist György Buzsáki of New York University goes one step further, arguing that if something is boosting faint signals to threshold in the brain, it is unlikely to be noise. 'Generating noise is very expensive,' he says. 'A good system, such as we presume the brain is, can't afford it.'

Buzsáki agrees with Miesenböck that there is probably a noise-like signal which modulates brain activity in mammals, but says there is no need to invoke specialised noise-making circuitry. Instead, he points to spontaneous neural activity occurring across the brain.

Neurons are capable of two types of activity, spontaneous and evoked. The first happens independently of an external stimulus, whereas the second is a response to it. Spontaneous activity is interesting to neuroscientists because it provides a possible mechanism for generating higher mental activity in the human brain. Spontaneous activity can spread over networks of neurons, and transient periods of synchronised neural firing at a rate of about 40 'spikes' a second. So-called gamma waves have been proposed as a way that different cognitive processes can be bound together to give rise to perception, for example.

Buzsáki says that faint incoming signals could piggyback on these spontaneous waves of activity and thus be lifted above threshold. This would be a more cost-effective way of enhancing a weak signal, he says, since spontaneous activity consumes little energy.

There is, of course, one key similarity between these two possibilities: both involve a signal that pushes another signal over a threshold. 'The principle is the same,' says Miesenböck. But the details matter both from the perspective of understanding the basic workings of the brain and, potentially, in order for us to exploit random noise and the phenomenon of stochastic resonance in future sensory aids such as retinal implants.

We will have to wait a little longer to find out whether natural selection created a brain with a built-in random noise generator, or simply one which is able to borrow some other neural signal to use as noise. Either way it seems that fly brains can't function without a little bit of dither – and that ours are probably dithering too.

The arbitrary ape

There's another way to make an evolutionary advantage out of randomness: think and behave in unexpected ways. This may even be the root of human creativity. Here's Dylan Evans.

We all have something of the Greek god in us. Proteus to be precise, who outwitted his enemies by constantly changing his shape. Humans may not go as far as transmogrification but when it comes to confusing a rival, our talent for erratic behaviour is second to none.

A rabbit pursued by a fox will bob and weave in a chaotic zigzag, rather than make a beeline for cover. Other animals use different forms of random behaviour to evade predators or catch their prey. But humans are the only ones who rely on unpredictability as a weapon in competition against each other, whether it be in a game of football or in international diplomacy.

Such behaviour has long been ignored, but researchers have woken up to the fact that not only can we behave in very random ways, but that such actions are far from pointless. Unpredictable behaviour may have evolved as a way of keeping our rivals in the dark. This could explain some of our strangest behaviour, such as sudden mood swings, and it also adds a whole new dimension to understanding the evolution of human intelligence. Astonishingly, our highly developed sense of the erratic may even be the spark that allows an ape adapted for savannah living to paint the Sistine Chapel, design the space shuttle and invent advertising slogans.

British biologist Michael Chance coined the phrase 'protean behaviour' in 1959, while at the University of Birmingham. But the evolutionary explanation for this phenomenon is somewhat more recent. It began with the observation by two British ethologists, Peter Driver and David Humphries, that many animals develop cognitive capacities so that they can predict the actions of their competitors or prey. Natural selection then favours mechanisms that make these actions harder to predict, so their enemies evolve better predictive powers, and an evolutionary arms race develops.

Two obvious ways of making your actions harder to predict are hiding your real intentions and giving out false signals. Both of these, however, are still vulnerable to the evolution of even better perceptual mechanisms on the part of the enemy, and so are not evolutionarily stable strategies – in other words, the arms race continues. In many conflicts the only way to stop this escalation is to adopt what game theorists call a 'mixed strategy', which bases decisions on probability. No amount of predictive talent will then prevail.

Submarine commanders in the Second World War hit on this idea and resorted to throwing dice to choose random patrol routes and so evade destroyers. In nature, interactions between enemies often work in a similar way. Sand eels, for example, usually react to predators by bunching together and swimming in a fast-moving school. But when threatened in a narrow pool, they behave very differently – the school breaks up and each eel darts about in random directions in an attempt to confuse the predator.

Driver and Humphries realised that protean behaviour should be common because of the competitive edge it gives species. Once they began looking, they found examples everywhere. There was the mobbing behaviour of gulls, which dive-bomb intruders from all directions to try to protect nesting colonies. And the herds of impala that burst into a whirlpool of activity, racing and plunging in every direction when disturbed.

Proteanism could also make sense of some of the more bizarre interactions between predators and prey. Many birds feign injury to lure the enemy away from a nest full of fledglings, using random changes in speed and direction to balance their aims of drawing attention away from their young and ensuring their own survival. Another puzzle – why moths, lizards and mice have mock convulsions when attacked – makes sense as a way of throwing a predator off its stride.

Competitive situations also bring out the Proteus in humans. But when biologists looked at people, they noticed an important difference between us and other animals – our competitors tend to be other humans. Geoffrey Miller, a psychologist at the University of New Mexico, Albuquerque, has highlighted this and suggested that this refinement in behaviour in our ancestors is key to our unique cognitive style. Our talent for thinking randomly may even be a source of the creative flair that sets humans apart from other animals.

Miller's ideas build on the theory of Machiavellian intelligence, which proposes that the main driving force in the evolution of human intelligence was the need to predict and manipulate the behaviour of other humans. The special cognitive capacities that evolved to deal with the social environment have been dubbed social intelligence. This includes calculated deception and its detection, but not protean behaviour. Miller argues that, in common with many other animals, our monkey-like ancestors had a basic ability to act randomly that they evolved to outwit predators. But during the transition from monkeys to apes to early hominids, this protean capacity was boosted by positive feedback from social intelligence, as outwitting our fellow humans became more important than outwitting other animals. As a result, he claims, proteanism plays a pivotal role in social intelligence.

Miller gives the following example to illustrate why protean behaviour would have evolved. Suppose our ancestors could have adopted one of two strategies for setting their anger threshold – the point at which they lose their temper. In the 'Old Faithful' strategy, the anger threshold is fixed. Those who adopt this strategy get angry only if an insult exceeds some predetermined level of annoyance. In the 'Mad Dog' strategy, on the other hand, the anger threshold varies randomly. Sometimes a big insult does not generate a response, but sometimes a small insult does. Which strategy would have been more effective?

If you are using Old Faithful, others quickly learn what they can get away with, so they constantly push you to the limit. But against the Mad Dog strategy any insult, however slight, might trigger retaliation. Furthermore, the person using this strategy does not have to waste time and effort punishing every small insult, because the uncertainty does most of the work. Flare up for no apparent reason every now and then, and people will tend to tiptoe around you. So Mad Dog is a much more effective way of outwitting your competitors.

'This might shed some light on the otherwise inexplicable nature of moods,' says Miller. When people explode over a minor insult that they would normally have laughed off, we assume that some particular event has triggered their bad mood. Miller, however, suggests that some moods may not be caused by any specific stimulus. 'They may simply be random alterations of our emotional state,' he says. 'The tendency to have such random mood changes could be a form of protean behaviour that evolved to make us less predictable and so less easy to exploit.'

But are we really natural born randomisers? For most of the last century, psychologists generally considered humans incapable of truly random behaviour. Dozens of studies seemed to confirm the view that producing a random series of responses is difficult, if not impossible, for humans. But most of these experiments involved placing people in artificial, non-competitive situations. Often, the researcher simply asked an isolated subject to write down a series of numbers with an instruction such as 'be as random as possible'. If proteanism in humans evolved as a way of outwitting other humans, as Miller argues, then people's failure to generate random numbers in these situations is not surprising. 'Psychologists failed to tap into our natural randomising abilities because they didn't expose subjects to the social games where those abilities evolved,' says Miller.

So in 1992, two Israeli psychologists set out to test people in face-to-face competition where there would be a motivation to randomise. David Budescu of Fordham University in New York State and Amnon Rapoport of the University of California, Riverside, got people to play a game called matching pennies. The rules are simple. Two players start with an equal number of coins. At each turn, both players simultaneously place a coin on the table between them. If the coins match (heads–heads or tails–tails) player A keeps both coins; if not, player B keeps them.

Though the players have opposite objectives, they both benefit from being able to predict what the other person will do next, and from making their own moves hard to predict. Mathematically, the best strategy is to pick heads and tails with equal probability, in a truly random series. Then over a long period of play, your contestant cannot gain the advantage. And this is exactly what Budescu and Rapoport found. The sequences of heads and tails generated by the two players came very close to true mathematical randomness, even though the players were given no instructions to that effect.

Another indication that randomness is an innate ability comes from the work of Allen Neuringer of Reed College in Portland, Oregon. He has shown that humans can learn to generate random sequences when given feedback. In one experiment, Neuringer asked students to generate a random series of a hundred pairs of 1s and 2s at a computer terminal. He then told the students how well they had done, measuring their performance by whether, for example, the series included approximately equal amounts of 1–1, 1–2, 2–1 and 2–2. In the first trial, the series was always non-random, but after several trials, the students' performances improved to the point that their series could not be distinguished from those generated by a computer.

A rat can learn to press a lever if you give it food as a reward, so is it surprising that students learn to generate random numbers? Yes, says Miller. The rat's behaviour is an example of conditioning – give it the right feedback and it will learn a new trick. But conditioning works by gradually eliminating random variation. 'It could never reinforce randomness itself,' he says. This leads him to conclude that there must be some innate randomising mechanism built into the mind. 'A roulette wheel in the head' is the metaphor used by the late John Maynard Smith of the University of Sussex. All sorts of processes can generate effectively random series, he has pointed out, so there is nothing bizarre about the idea that the brain might be able to do so.

Many animals seem to have this mental roulette wheel but, Miller argues, by refining its abilities humans have developed a mechanism that is capable of more than simply outwitting enemies. Our super-protean capacity is the basis for our inventiveness and artistic creativity, he says. 'Proteanism provides a key element of creativity that other mental mechanisms lack – the capacity for rapid, unpredictable generation of highly variable alternatives,' says Miller. Studies of human creativity often emphasise this element. Without it, for example, there would be no brainstorming. And in many forms of art, from music to comedy, coming up with a new twist on an old theme or confounding an audience's expectations is the key to success.

The prevailing view is that human creativity came about as a lucky accident, through the increasing overlap of cognitive capacities designed for other functions. Ecological intelligence evolved to meet the complex demands of foraging for food in the savannah, technical intelligence developed with our tool-making skills, and social intelligence with group living. Steven Mithen, an archaeologist at the University of Reading, has argued that in the early hominid mind these intellectual specialities were walled off from each other like the chapels of an early cathedral. He claims that the modern mind evolved only with the collapse of these mental divisions and the development of more general cognitive capacities.

The problem with this view, says Miller, is that it is at odds with one of the main features of natural selection – that it tends to lead to increased specialisation rather than increased generalisation. Miller's theory, however, requires no appeal to increasingly general mechanisms. On the contrary, an innate randomising mechanism could well be a very specialised way of generating novel ideas. Miller speculates that it might work by amplifying the quantum mechanical noise in synaptic activity. Alternatively, it could work in the same way that computers generate random numbers: producing 'pseudo-randomness' by feeding the numbers it generates back into a program that is too complex to be worked out by an outsider.

According to the Machiavellian intelligence hypothesis, creativity is a spin-off from social intelligence alone. The idea is that our ancestors first evolved to cope with savannah life, then learnt to exploit their environment using tools, and finally perfected the art of social living. It was only then that creativity really took off. But until now, nobody has come up with a plausible explanation of how this might have happened. Miller's theory could have the answer by showing how proteanism evolved in the social setting, and then making the link between randomness and creativity.

Evolutionary theorists have tended to see evolutionary adaptation as a process that increases order and complexity. Natural selection was thought to build improbable regularities from random disorder. Protean behaviour defies this simple view – it is at once random and adaptive, chaotic and yet the result of selection. No wonder it took biologists so long to see it.

6

Putting chance to work

We have arrived at our destination: hopefully we now understand randomness, and know its limitations and usefulness in the world around us. It's time to harness all that knowledge, and see if we can use chance to enhance our lives.

It's a techno-lottery

First in our look at applications of randomness is the problem and opportunity of generating random numbers. It's that Michael Brooks again.

Mads Haahr is in no doubt. 'Generating randomness is not a task that should be left to humans,' he says.

You might expect him to say that. A computer scientist at Trinity College Dublin, he is the creator of a popular online random number generator, hosted at www.random.org. But he has a point. Human brains are wired to spot and generate patterns. That is useful when it's all about seeing predators on the savannah before they see you, but it handicaps us when we need to think in random and unpredictable ways. Though we can learn to randomise to a certain extent, our brains can only go so far. It's a shame, Haahr says: true randomness is a useful thing to have at your disposal.

Random numbers are used in cryptography, computing, design and many other applications. Our inability to 'do' random properly means that we usually have to outsource it to machines. The thing is, relying on outside sources of randomness has its own problems. The first dice for divination and gaming, for instance, were six-sided bones from the heels of sheep, with numbers carved into the faces. The shape made some numbers more likely to appear than others, giving a decisive advantage to those who understood its properties.

Suspicion about the reliability of randomness generators remains with modern equivalents like casino dice, roulette wheels or lottery balls. But it is online where it really matters. Generating random strings of numbers is essential not just for gambling games or shuffling songs on your iPod, but also to produce unguessable keys used to encrypt sensitive digital information. 'I don't think people are very conscious of how important randomness is for the security of their data,' says Haahr.

And it takes more than programming. You can't just give computers rules to create random numbers; that wouldn't be random. Instead you might use an algorithm to 'seed' a random-looking output from a smaller, unpredictable input: use the date and time to determine which random digits to extract from a random number string such as pi, say, and work from there. The problem is that such 'pseudo-random' numbers are limited by the input, and tend to repeat non-randomly after a certain time in a way that is guessable if you see enough of them.

An alternative is to hook up your computer to some source of physical, 'true' randomness. In the 1950s, the UK Post Office wanted a way to generate industrial quantities of random numbers to pick the winners of its Premium Bonds lottery. The job fell to the designers of the pioneering Colossus computer, developed to crack Nazi Germany's wartime codes. They created ERNIE, the Electronic Random Number Indicator Equipment, which harnessed the chaotic trajectories of electrons passing through neon tubes to produce a randomly timed series of electronic pulses that seeded a random number.

ERNIE is now in his fourth iteration and is a simpler soul, relying on thermal noise from transistors to generate randomness. Many modern computing applications use a similar source, collected using on-chip generating units such as Intel's RdRand and Via's Padlock. Haahr's generator takes its seed from intrinsically noisy atmospheric processes.

Two problems remain. First, with enough computing power anyone can, in theory, reconstruct the processes of classical physics that created the random numbers. Second, and more practically, random number generators based solely on physical processes often can't produce random bits fast enough.

Many systems, such as the Unix-based platforms used by Apple, get round the first problem by combining the output from on-chip randomness generators with the contents of an 'entropy pool', filled with other random contributions. This could be anything from thermal noise in devices connected to the computer to the random timings of the user's keyboard strokes. The components are then combined using a 'hash function' to generate a single random number. Hash functions are the mathematical equivalent of stirring ink into water: there's no known way to work out what the set of inputs was, given the number the function spits out. However, that doesn't mean there couldn't be in the future – and there's still the speed problem. Here, the workaround is generally to use a physical random number generator only as a seed for a program that generates a more abundant flow.

But there we are back with the algorithm problem. The precise nature of the methods these programs use is proprietary, but in 2013 security analysts raised concerns that the US National Security Agency knew the internal workings of one such generator, called Dual_EC_DRBG, potentially allowing them to break encryptions that relied on it. If you're just playing online games, that's not a big problem. But when making multibillion-dollar financial transactions, or encrypting sensitive documents, a suspicion that you are being watched is a bigger deal.

Such difficulties lead some researchers to suggest that we'll never have an uncrackable source of randomness as long as we rely on the classical world, where randomness is not intrinsic, but comes down to who has what information. For safer encryption, we must turn to quantum physics, where things truly do seem random. Instead of a coin toss, you might ask whether a photon hitting a half-silvered mirror passed through it or was reflected. Instead of rolling a dice, you might present an electron with a choice of six circuits to pass through. 'As a mathematician, I like my randomness to come with proof, and quantum random numbers give us that,' says Carl Miller of the University of Michigan in Ann Arbor. 'It's unique in that respect.'

Cryptographic systems that exploit the vagaries of quantum theory for more secure communication do exist. But even they are not the last word in security. Extracting quantum randomness always involves someone making non-random choices about equipment, measurements and suchlike. The less-than-perfect efficiency of photon detectors used in some methods could also provide a back door through which non-randomness can slip in.

One way out that is still under investigation might be to amplify quantum randomness so that you always have more of it than anyone can hack. Ways exist in theory to turn n random bits into 2^n bits of pure randomness, and also to launder bits to remove any correlation with the device that first made them.

Such device-independent quantum random number generation is just the latest development in our search for true randomness. Chances are, this too will soon become reality – only then for someone to find a way to game it. With humans forever in the mix, it could be that we'll always be searching for randomness we can rely on.

Locating, locating, locating

If you've ever played hide-and-seek, the chances are you invoked a little touch of randomness. It's surprisingly efficient to trust that you'll find something by chance – another evolved trick where randomness plays a surprising role, as Kate Ravilious explains.

Under the bed? Behind the cushions? In a forgotten pocket? Searching for lost keys usually feels like a random hunt – or so you might think. While frantically overturning household items might seem a pot-luck approach, it now appears that the apparently chaotic way we look for things might reflect a method perfected by our hunter-gatherer ancestors over millennia of evolution. It's a realisation that could have surprisingly profound effects in fields as diverse as anthropology, town planning, archaeology and – believe it or not – learning how to survive a trip to Ikea.

Searching has always been crucial to human survival. Hunter-gatherers had to be good at searching to find food and water. What's more, their movement had knock-on effects on many phenomena, such as the spread of populations, the advance of disease and the emergence of civilisations. So modelling the migration of ancient hunter-gatherer communities helps us understand all these things.

Traditionally, scientists have assumed that ancient peoples moved from place to place in a random way. They based their models on a description of random movement borrowed from physics, called Brownian motion, which can accurately describe many different diffusion situations, such as the movement of ink blotting through paper, smoke in the air or pollen grains floating on the surface of a pond.

In Brownian motion the probability of taking a step of a particular length follows what statisticians call a normal distribution, meaning that there is a higher probability that the particle will take short- to medium-length steps and a much lower probability that it will take very long steps. Although no one had proved that ancient tribes moved in this way, no one had disproved it either, and it was widely accepted. The same was assumed to hold for the way animals and insects forage for food.

However, it turns out that many animals – including bumblebees, albatrosses, jackals, reindeer, spider monkeys and even zooplankton – don't forage in a Brownian pattern at all. Instead, their movement matches a model mathematicians call a Lévy flight, named after French mathematician Paul Lévy. This describes a special kind of random motion comprised of short jumps that cluster in a small area interspersed with long leaps to a new area.

In a Lévy flight, the probability of taking a step of a particular length follows a power law distribution, meaning that very short steps and very long steps are more likely than in Brownian motion, and medium-length steps are less likely.

Lévy flights are the optimum way of foraging for food in the natural world, says Bruce West, a physicist at the Army Research Office in North Carolina, who studies Lévy flight patterns in the natural world. 'A Lévy flight strategy means you avoid going back to the places that are depleted of resources.'

Clifford Brown, an archaeologist from Florida Atlantic University at Boca Raton, noticed these revelations and wondered if they might apply to humans too. Brown has a long-standing interest in fractal patterns that occur in nature and had come across Lévy flights before in connection with natural fractal phenomena. To discover whether Lévy flights held for human migration, he decided to return to first principles and began searching for concrete data on human movement.

No empirical data exists on the movement of ancient hunter-gatherer tribes, so Brown chose something nearly as good: a set of intimately detailed records of the movements of one of the world's few remaining hunter-gatherer tribes, the Dobe Ju/'hoansi Bushmen (also known as the !Kung). The Bushmen have inhabited the Dobe area of the Kalahari Desert, straddling the border of Namibia and Botswana, for thousands of years. Although most have now resettled, up until the late 1960s they lived in the traditional way, gaining as much as 85 per cent of their diet from hunting and gathering.

In 1968 John Yellen, an anthropologist at the Smithsonian Institution in Washington DC, spent six months living with the Ju/'hoansi, recording their lifestyle, how far they travelled and how much time they spent in each place as they hunted and moved from camp to camp.

Brown used Yellen's notes to piece together a map of how the Ju/'hoansi had moved around. In just six months the tribe moved 37 times and set up 28 different camps. At first glance, the pattern of movements looked haphazard, with frantic searching for food and water in one area, followed by long treks to new places. But upon careful analysis Brown discovered a distinctive pattern behind their movement – sure enough, the probability distribution of the distance moved each time and the time spent in each camp fitted a Lévy flight model almost perfectly. It looks like the Ju/'hoansi moved in Lévy steps because it brought some distinct advantage when searching for food, Brown says. So what might that be?

The Kalahari is a harsh environment where water and food are scarce. One of the most important food sources is the mongongo tree, whose nuts are a major source of nutrition. The trees tend to grow along the crests of ancient east-west sand dunes, with water holes dotted about in between. Brown noticed that the mongongo trees were distributed in tightly packed clusters separated by large areas devoid of the trees. The Ju/'hoansi seem to have learned to move in a Lévy flight pattern as a result of the distribution of this food resource, says Brown.

He also believes the Ju/'hoansi were not the only people to have adopted Lévy flights. 'Many natural food resources are distributed in fractal patterns, such that exploiting them would tend to move people around in a Lévy flight pattern,' he says.

So if people moved in Lévy flights when hunting and foraging, might they move in Lévy flights when exploring? Brown thinks it certainly merits further investigation. Michael Rosenberg, a computational evolutionary biologist from Arizona State University in Tempe, agrees that it is time to rethink our old models of early human migration. 'This evidence provides us with justification to try Lévy flight models,' he says.

Already, other examples are emerging that support Brown's observations. Marek Zvelebil, who worked as an archaeologist at the University of Sheffield in the UK until his death in 2011, analysed the advance of farming across

Europe and noticed a pattern of motion that includes what he describes as 'leapfrogging' from one region to another. This led him to conclude that farming communities must have sent out pioneers to identify new areas to settle – rather than spreading out slowly from established territory as was previously assumed. He suggested a Lévy flight pattern might fit the spread of these peoples.

Not everyone agrees that early farmers so closely match the hunter-gatherers, however. 'For hunter-gatherers it is not difficult to up sticks and move,' says Graeme Ackland, a physicist at the University of Edinburgh, UK, who has been modelling the advance of farming populations. Ackland thinks it is unlikely these communities moved in the same way. His simulations used Brownian motion driven by population pressure to model how early agricultural peoples moved across Europe, and found they advanced at around 1 kilometre per year. Brown accepts that Lévy flights may not be the answer to every kind of human movement. 'It may not apply if people have a different diet and distribution of resources,' he concedes.

All the same, Lévy flights might explain one of the biggest migration puzzles of all time: how prehistoric native Americans settled the New World. Arriving over the Bering land bridge from Siberia to Alaska around 11,500 years ago, when sea levels were much lower, the so-called Clovis people radiated southwards, reaching the southern tip of South America in just 1,000 years. They were a hunter-gatherer tribe living off mammoth and other game, which they hunted using distinctive fluted stone spear-points. No one has yet explained how and why these people travelled many thousands of kilometres in such a short time. 'Lévy flights give an explanation of how these people could have moved so fast,' says Brown.

Meanwhile, Rosenberg thinks that the Lévy flight model could even reveal the first steps of humankind. 'I think it would be interesting to see if it could help to explain some of the huge jumps that we see in the "out of Africa" hypothesis of human evolution,' he says.

And there are other conundrums that look like they might yield to a Lévy flight explanation. For example, the spread of the hereditary disease sickle-cell anaemia through central Africa took just a few thousand years, much faster than the tens of thousands of years we would expect had it followed random Brownian motion. Sickle-cell anaemia is associated with the spread of malaria, which in turn is associated with the spread of farming. If sickle-cell anaemia spread fast, then both farming and malaria must have moved rapidly too. 'Lévy flight models could help us to describe this gene flow,' says Henry Harpending, an anthropologist at the University of Utah in Salt Lake City who studies the migration of ancient African populations.

Could Lévy flights still be influencing the way we live now? 'It's certainly a possibility,' says Brown. Others agree. Alan Penn, an architect from University College London, is applying this philosophy to the design of new towns and cities. Penn and his colleagues analysed the layout of city shops and showed that they tend to form a blotchy pattern resembling a Lévy flight distribution. 'Shops that are similar tend to group together. This means they compete, but they also attract the crowds,' says Penn.

London is a perfect example of this. Tottenham Court

Road is home to a cluster of electrical goods shops, Hatton Garden is the place for jewels, and Cork Street is where to go for fine art. In each of these areas there is a main street that acts like an artery carrying the flow of people, plus back streets where shoppers can poke around. On a smaller scale you can observe the same pattern in a market, with fruit and veg stalls in one corner, fish lining another alley and meat in yet another.

Penn created a computer model of a town into which he placed 'agents' – representing people. The agents were programmed to search for goods by moving three steps at a time in a random direction within their field of view. If enough agents demanding a particular product congregated in one place, the computer would create a shop there selling what the agents wanted. Penn found that small clusters of similar shops would emerge over time. 'It seemed to help the agents be more efficient in finding the shops,' he says. This implies that the layout of our towns and cities may well have been shaped by the search patterns we learned from our hunter-gatherer past.

Penn is now using his models to help urban planners rejuvenate ailing areas. One of his projects has been to help revitalise London's South Bank cultural quarter, where he suggested introducing shortcuts between key locations and clustering sites of interest such as cafes, restaurants and bookshops. Shops are easier to find if they are distributed in a pattern that mirrors the natural way we look for things, says Penn. He is also using these methods to help develop more pedestrian-friendly living areas in Milton Keynes, and other places in the south-east of England.

Brown plans to use his new understanding of Lévy

flights to help identify archaeologically interesting sites. 'In many parts of the world hunter-gatherers occupied the landscape for very long periods of time, but finding evidence of their campsites and movements is hard,' he says. Conventional searches consist of choosing a region and then digging sample sites at equal intervals, which can be very expensive and time-consuming. Brown thinks that guiding the sampling according to a Lévy flight pattern could cut costs and speed up search times. Ironically, the search patterns our long-lost ancestors bequeathed to us might be our best bet for rediscovering them.

I, algorithm

We have already seen that computers are not great at generating randomness, but it turns out that they can use chance to their advantage. Artificial intelligence – machines with chance at their heart – is finally coming of age, and finding applications in areas as diverse as the oversight of nuclear explosions and preserving the fragile health of a premature baby. Anil Ananthaswamy tells us how.

Given the choice between a flesh-and-blood doctor and an artificial intelligence system for diagnosing diseases, Pedro Domingos is willing to stake his life on AI. 'I'd trust the machine more than I'd trust the doctor,' says Domingos, a computer scientist at the University of Washington, Seattle. Considering the bad rap AI usually receives – overhyped, underwhelming – such strong statements in its support are rare indeed.

Back in the 1960s, AI systems started to show great promise for replicating key aspects of the human mind. Scientists began by using mathematical logic to both represent knowledge about the real world and to reason about it, but it soon turned out to be an AI straitjacket. While logic was capable of being productive in ways similar to the human mind, it was inherently unsuited to dealing with uncertainty.

Yet after spending so long shrouded in a self-inflicted winter of discontent, the much-maligned field of AI is in bloom again. Sophisticated computer systems are beginning to exhibit the nascent abilities which sparked interest in AI in the first place: the ability to reason like humans, even in a noisy and chaotic world.

Lying close to the heart of AI's revival is a technique called probabilistic programming, which combines the logical underpinnings of the old AI with the power of statistics and probability. 'It's a natural unification of two of the most powerful theories that have been developed to understand the world and reason about it,' says Stuart Russell, a pioneer of modern AI at the University of California, Berkeley. This powerful combination is finally starting to disperse the fog of the long AI winter. 'It's definitely spring,' says cognitive scientist Josh Tenenbaum at the Massachusetts Institute of Technology.

The term 'artificial intelligence' was coined in 1956 by John McCarthy of MIT. At the time, he advocated the use of logic for developing computer systems capable of reasoning. This approach matured with the use of so-called first-order logic, in which knowledge about the real world is modelled using formal mathematical symbols and

notations. It was designed for a world of objects and rela-
tions between objects, and it could be used to reason about
the world and arrive at useful conclusions. For example, if
person X has disease Y, which is highly infectious, and X
came in close contact with person Z, using logic one can
infer that Z has disease Y.

However, the biggest triumph of first-order logic was
that it allowed models of increasing complexity to be
built from the smallest of building blocks. For instance,
the scenario above could easily be extended to model the
epidemiology of deadly infectious diseases and draw
conclusions about their progression. Logic's ability to
compose ever-larger concepts from humble ones even
suggested that something analogous might be going on
in the human mind.

That was the good news. 'The sad part was that,
ultimately, it didn't live up to expectations,' says Noah
Goodman, a cognitive scientist at Stanford University in
California. That's because using logic to represent knowledge, and reason about it, requires us to be precise in our
know-how of the real world. There's no place for ambigu-
ity. Something is either true or false, there is no maybe.
The real world, unfortunately, is full of uncertainty, noise
and exceptions to almost every general rule. AI systems
built using first-order logic simply failed to deal with it.
Say you want to tell whether person Z has disease Y. The
rule has to be unambiguous: if Z came into contact with
X, then Z has disease Y. First-order logic cannot handle a
scenario in which Z may or may not have been infected.

There was another serious problem: it didn't work
backwards. For example, if you knew that Z had disease

Y, it was not possible to infer with absolute certainty that Z caught it from X. This typifies the problems faced in medical diagnosis systems. Logical rules can link diseases to symptoms, but a doctor faced with symptoms has to infer backwards to the cause. 'That requires turning around the logic formula, and deductive logic is not a very good way to do that,' says Tenenbaum.

These problems meant that by the mid-1980s, the AI winter had set in. In popular perception, AI was going nowhere. Yet Goodman believes that, secretly, people didn't give up on it. 'It went underground,' he says.

The first glimmer of spring came with the arrival of neural networks in the late 1980s. The idea was stunning in its simplicity. Developments in neuroscience had led to simple models of neurons. Coupled with advances in algorithms, this let researchers build artificial neural networks (ANNs) that could learn, ostensibly like a real brain. Invigorated computer scientists began to dream of ANNs with billions or trillions of neurons. Yet it soon became clear that our models of neurons were too simplistic and researchers couldn't tell which of the neuron's properties were important, let alone model them.

Neural networks, however, helped lay some of the foundations for a new AI. Some researchers working on ANNs eventually realised that these networks could be thought of as representing the world in terms of statistics and probability. Rather than talking about synapses and spikes, they spoke of parameterisation and random variables. 'It now sounded like a big probabilistic model instead of a big brain,' says Tenenbaum.

Then, in 1988, Judea Pearl at the University of

California, Los Angeles, wrote a landmark book called *Probabilistic Reasoning in Intelligent Systems*, which detailed an entirely new approach to AI. Behind it was the theorem developed in the 18th century by Thomas Bayes which links the conditional probability of an event P occurring given that Q has occurred to the conditional probability of Q given P. Here was a way to go back-and-forth between cause and effect. 'If you can describe your knowledge in that way for all the different things you are interested in, then the mathematics of Bayesian inference tells you how to observe the effects, and work backwards to the probability of the different causes,' says Tenenbaum.

The key is a Bayesian network, a model made of various random variables, each with a probability distribution that depends on every other variable. Tweak the value of one, and you alter the probability distribution of all the others. Given the value of one or more variables, the Bayesian network allows you to infer the probability distribution of other variables – in other words, their likely values. Say these variables represent symptoms, diseases and test results. Given test results (a viral infection) and symptoms (fever and cough), one can assign probabilities to the likely underlying cause (flu, very likely; pneumonia, unlikely).

By the mid-1990s, researchers including Russell began to develop algorithms for Bayesian networks that could utilise and learn from existing data. In much the same way as human learning builds strongly on prior understanding, these new algorithms could learn much more complex and accurate models from much less data. This was a huge step up from ANNs, which did not allow for prior knowledge; they could only learn from scratch for each new problem.

The pieces were falling into place to create an artificial intelligence for the real world. The parameters of a Bayesian network are probability distributions, and the more knowledge one has about the world, the more useful these distributions become. But unlike systems built with first-order logic, things don't come crashing down in the face of incomplete knowledge.

Logic, however, was not going away. It turns out that Bayesian networks aren't enough by themselves because they don't allow you to build arbitrarily complex constructions out of simple pieces. Instead it is the synthesis of logic programming and Bayesian networks into the field of probabilistic programming that is creating a buzz.

At the forefront of this new AI are a handful of computer languages that incorporate both elements. There's Church, developed by Goodman, Tenenbaum and colleagues, and named after Alonzo Church, who pioneered a form of logic for computer programming. Then there's Markov Logic Network, developed by Domingos's team and combining Markov networks – similar to Bayesian networks – with logic. Russell and his colleagues have the straightforwardly named Bayesian Logic (BLOG).

Russell has demonstrated the expressive power of such languages in places such as the headquarters of the UN's Comprehensive Test Ban Treaty Organization (CTBTO) in Vienna, Austria. The CTBTO invited Russell on a hunch that the new AI techniques might help with the problem of detecting nuclear explosions. After a morning listening to presenters speak about the challenge of detecting the seismic signatures of far-off nuclear explosions amidst the background of earthquakes, the vagaries of signal propagation through the Earth, and noisy detectors at seismic stations worldwide, Russell sat down to model the problem using probabilistic programming. 'And in the lunch hour I was able to write a complete model of the whole thing,' says Russell. It was half a page long.

Prior knowledge can be incorporated into this kind of model, such as the probability of an earthquake occurring in Sumatra, Indonesia, versus Birmingham, UK. The CTBTO also requires that any system assumes that a nuclear detonation occurs with equal probability anywhere on Earth. Then there is real data – signals received at CTBTO's monitoring stations. The job of the AI system is to take all of this data and infer the most likely explanation for each set of signals.

Therein lies the challenge. Languages like BLOG are equipped with so-called generic inference engines. Given a model of some real-world problem, with a host of variables and probability distributions, the inference engine has to calculate the likelihood of, say, a nuclear explosion in the Middle East, given prior probabilities of expected events and new seismic data. But change the variables to represent symptoms and disease and it then must be capable of medical diagnosis. In other words its algorithms must be very general. That means they will be extremely inefficient.

The result is that these algorithms have to be customised for each new challenge. But you can't hire a PhD student to improve the algorithm every time a new problem comes along, says Russell. 'That's not how your brain works; your brain just gets on with it.'

This is what gives Russell, Tenenbaum and others pause, as they contemplate the future of AI. 'I want people to be excited but not feel as if we are selling snake oil,' says Russell. Tenenbaum agrees; even as a relatively young researcher, he thinks there is only a 50:50 chance that the challenge of efficient inference will be met in his lifetime. And that's despite the fact that computers will get faster and algorithms smarter. 'These problems are much harder than getting to the moon or Mars,' he says.

This, however, is not dampening the spirits of the AI community. Daphne Koller of Stanford University, for instance, is attacking very specific problems using probabilistic programming and has much to show for it. Along with neonatologist Anna Penn, also at Stanford, and colleagues, Koller has developed a system called PhysiScore for predicting whether a premature baby will have any health problems – a notoriously difficult task. Doctors are unable to predict this with any certainty, 'which is the only thing that matters to the family', says Penn.

PhysiScore takes into account factors such as gestational age and weight at birth, along with real-time data collected in the hours after birth, including heart rate, respiratory rate and oxygen saturation. 'We are able to tell within the first 3 hours which babies are likely to be healthy and which are much more likely to suffer severe complications, even if the complications manifest after 2 weeks,' says Koller.

'Neonatologists are excited about PhysiScore,' says Penn. As a doctor, Penn is especially pleased about the ability of AI systems to deal with hundreds, if not thousands, of variables while making a decision. This could make them even better than their human counterparts. 'These tools make sense of signals in the data that we doctors and nurses can't even see,' says Penn.

This is why Domingos places such faith in automated medical diagnosis. One of the best known is the Quick Medical Reference, Decision Theoretic (QMR-DT), a Bayesian network which models hundreds of significant diseases and thousands of related symptoms. Its goal is to infer a probability distribution for diseases given some symptoms. Researchers have fine-tuned the inference algorithms of QMR-DT for specific diseases, and taught it using patients' records. 'People have done comparisons of these systems with human doctors and the [systems] tend to win,' says Domingos. 'Humans are very inconsistent in their judgements, including diagnosis. The only reason these systems aren't more widely used is that doctors don't want to let go of the interesting parts of their jobs.'

There are other successes for such techniques in AI, one of the most notable being speech recognition, which has gone from being laughably error-prone to impressively precise. Doctors can now dictate patient records and speech recognition software turns them into electronic documents, limiting the use of manual transcription. Language translation is also beginning to replicate the success of speech recognition.

But there are still areas that pose significant challenges. Understanding what a robot's camera is seeing is one. Solving this problem would go a long way towards creating robots that can navigate by themselves.

Besides developing inference algorithms that are flexible and fast, researchers must also improve the ability of AI systems to learn, whether from existing data or from the real world using sensors. Today, most machine learning is done by customised algorithms and carefully constructed data sets, tailored to teach a system to do something specific. 'We'd like to have systems that are much more versatile, so that you can put them in the real world, and they learn from a whole range of inputs,' says Koller.

The ultimate goal for AI, as always, is to build machines that replicate human intelligence, but in ways that we fully understand. 'That could be as far off, and maybe even as dangerous, as finding extra-terrestrial life,' says Tenenbaum. 'Human-like AI, which is a broader term, has room for modesty. We'd be happy if we could build a vision system which can take a single glance at a scene and tell us what's there – the way a human can.'

The power of one

Every now and then, a simple idea takes the world by storm. Benford's law is one such idea. All it involves is counting the different types of digits in a set of numbers, to see how randomly they are distributed. As Robert Matthews shows here, this simple idea has put people behind bars: you break Benford's law at your peril.

Alex had no idea what dark little secret he was about to uncover when he asked his brother-in-law to help him out with his term project. As an accountancy student at Saint Mary's University in Halifax, Nova Scotia, Alex needed some real-life commercial figures to work on, and his brother-in-law's hardware store seemed the obvious place to get them.

Trawling through the year's sales figures, Alex could find nothing obviously strange about them. Still, he did what he was supposed to do for his project, and performed a bizarre little ritual requested by his accountancy professor, Mark Nigrini. He went through the sales figures and made a note of how many started with the digit 1. It came out at 93 per cent. He handed it in and thought no more about it.

Later, when Nigrini was marking the coursework, he took one look at that figure and realised that an embarrassing situation was looming. His suspicions hardened as he looked through the rest of Alex's analysis of his brother-in-law's accounts. None of the sales figures began with the digits 2 through to 7, and there were just 4 beginning with the digit 8, and 21 with 9. After a few more checks, Nigrini was in no doubt: Alex's brother-in-law was a fraudster, systematically cooking the books to avoid the attentions of bank managers and tax inspectors.

It was a nice try. At first glance, the sales figures showed nothing very suspicious, with none of the sudden leaps or dives that often attract the attentions of the authorities. But that was just it: they were too regular. And this is why they fell foul of that ritual Nigrini had asked Alex to perform.

Because what Nigrini knew, and Alex's brother-in-law clearly didn't, was that the digits making up the shop's sales figures should have followed a mathematical rule discovered by accident over 100 years ago. Known as Benford's law, it is a rule obeyed by a stunning variety of phenomena, from stock market prices to census data to the heat capacities of chemicals. Even a ragbag of figures extracted from newspapers will obey the law's demands that around 30 per cent of the numbers will start with a 1, 18 per cent with a 2, right down to just 4.6 per cent starting with a 9.

It is a law so unexpected that at first many people simply refuse to believe it can be true. But after years of being regarded as a mathematical curiosity, Benford's law is now being eyed by everyone from forensic accountants to computer designers, all of whom think it could help them solve some tricky problems with astonishing ease.

The story behind the law's discovery is every bit as weird as the law itself. In 1881, the American astronomer Simon Newcomb penned a note to the *American Journal of Mathematics* concerning a strange quirk he'd noticed about books of logarithms, then widely used by scientists performing calculations. The first pages of such books seemed to get grubby much faster than the last ones.

The obvious explanation was perplexing. For some reason, people did more calculations involving numbers starting with 1 than with 8 and 9. Newcomb came up with a little formula that matched the pattern of use pretty well: nature seems to have a penchant for arranging numbers so that the proportion beginning with the digit D is equal to $\log_{10}$ of $1 + (1/D)$ (see 'Here, there and everywhere', page 231).

With no particularly convincing argument for why the

formula should work, Newcomb's paper failed to arouse much interest, and the Grubby Pages Effect was forgotten for over half a century. But in 1938, a physicist with the General Electric Company in the US, Frank Benford, rediscovered the effect and came up with the same law as Newcomb. But Benford went much further. Using more than 20,000 numbers culled from everything from listings of the drainage areas of rivers to numbers appearing in old magazine articles, Benford showed that they all followed the same basic law: around 30 per cent began with the digit 1, 18 per cent with 2 and so on.

Like Newcomb, Benford did not have any really good explanation for the existence of the law. Even so, the sheer wealth of evidence he provided to demonstrate its reality and ubiquity has led to his name being linked with the law ever since.

It was nearly a quarter of a century before anyone came up with a plausible answer to the central question: why on earth should the law apply to so many different sources of numbers? The first big step came in 1961 with some neat lateral thinking by Roger Pinkham, a mathematician then at Rutgers University in New Brunswick, New Jersey. Just suppose, said Pinkham, there really is a universal law governing the digits of numbers that describe natural phenomena such as the drainage areas of rivers and the properties of chemicals. Then any such law must work regardless of what units are used. Even the inhabitants of the Planet Zob, who measure area in grondekis, must find exactly the same distribution of digits in drainage areas as we do, using hectares. But how is this possible, if there are 87.331 hectares to the grondeki?

The answer, said Pinkham, lies in ensuring that the distribution of digits is unaffected by changes of units. Suppose you know the drainage area in hectares for a million different rivers. Translating each of these values into grondekis will change the individual numbers, certainly. But overall, the distribution of numbers would still have the same pattern as before. This is a property known as 'scale invariance'.

Pinkham showed mathematically that Benford's law is indeed scale-invariant. Crucially, however, he also showed that Benford's law is the only way to distribute digits that has this property. In other words, any 'law' of digit frequency with pretensions of universality has no choice but to be Benford's law.

Pinkham's work gave a major boost to the credibility of the law, and prompted others to start taking it seriously and thinking up possible applications. But a key question remained: just what kinds of numbers could be expected to follow Benford's law? Two rules of thumb quickly emerged. For a start, the sample of numbers should be big enough to give the predicted proportions a chance to assert themselves. Second, the numbers should be free of artificial limits, and allowed to take pretty much any value they please. It is clearly pointless expecting, say, the prices of ten different types of beer to conform to Benford's law. Not only is the sample too small, but more importantly the prices are forced to stay within a fixed, narrow range by market forces.

On the other hand, truly random numbers won't conform to Benford's law either: the proportions of leading digits in such numbers are, by definition, equal.

Benford's law applies to numbers occupying the 'middle ground' between the rigidly constrained and the utterly unfettered.

Precisely what this means remained a mystery until 1996, when mathematician Theodore Hill of Georgia Institute of Technology in Atlanta uncovered another insight into the origin of Benford's law. It comes, he realised, from the various ways in which different kinds of measurements tend to spread themselves. Ultimately, everything we can measure in the universe is the outcome of some process or other: the random jolts of atoms, say, or the exigencies of genetics. Mathematicians have long known that the spread of values for each of these follows some basic mathematical rule. The heights of bank managers, say, follow the bell-shaped Gaussian curve, daily temperatures rise and fall in a wave-like pattern, while the strength and frequency of earthquakes are linked by a logarithmic law.

Now imagine grabbing random handfuls of data from a hotchpotch of such distributions. Hill proved that as you grab ever more of such numbers, the digits of these numbers will conform ever closer to a single, very specific law. This law is a kind of ultimate distribution, the 'Distribution of Distributions'. And he showed that its mathematical form is ... Benford's law.

Hill's theorem goes a long way to explaining the astonishing ubiquity of Benford's law. For while numbers describing some phenomena are under the control of a single distribution such as the bell curve, many more, describing everything from census data to stock-market prices, are dictated by a random mix of all kinds of distributions. If Hill's theorem is correct, this means that the digits of these data should follow Benford's law. And, as Benford's own monumental study and many others have showed, they really do.

Mark Nigrini, Alex's former project supervisor and now a professor of accountancy at the College of New Jersey in Ewing, sees Hill's theorem as a crucial breakthrough: 'It ... helps explain why the significant-digit phenomenon appears in so many contexts.'

It has also helped Nigrini to convince others that Benford's law is much more than just a bit of mathematical frivolity. Over the past few years, Nigrini has become the driving force behind a far from frivolous use of the law: fraud detection.

In a ground-breaking doctoral thesis published in 1992, Nigrini showed that many key features of accounts, from sales figures to expenses claims, follow Benford's law, and that deviations from the law can be quickly detected using standard statistical tests. Nigrini calls the fraud-busting technique 'digital analysis', and its successes are starting to attract interest in the corporate world and beyond.

Some of the earliest cases, including the sharp practices of Alex's store-keeping brother-in-law, emerged from student projects set up by Nigrini. But soon he was using digital analysis to unmask much bigger frauds. One case involved an American leisure and travel company with a nationwide chain of motels. Using digital analysis, the company's audit director discovered something odd about the claims being made by the supervisor of the company's healthcare department. 'The first two digits of the health-care payments were checked for conformity to Benford's law, and this revealed a spike in numbers beginning with the digits '65',' says Nigrini. 'An audit showed 13 fraudulent cheques for between $6,500 and $6,599 … related to fraudulent heart surgery claims processed by the supervisor, with the cheque ending up in her hands.'

Benford's law had caught the supervisor out, despite her best efforts to make the claims look plausible. 'She carefully chose to make claims for employees at motels with a higher than normal number of older employees,' says Nigrini. 'The analysis also uncovered other fraudulent claims worth around $1 million in total.'

Not surprisingly, big businesses and central governments are now also taking Benford's law seriously. 'Digital analysis is being used by listed companies, large private companies, professional firms and government agencies in the US and Europe, and by one of the world's biggest audit firms,' says Nigrini.

The technique is also attracting interest from those hunting for other kinds of fraud. At the International Institute for Drug Development in Brussels, Marc Buyse and his colleagues believe Benford's law can reveal suspicious data in clinical trials, while a number of university researchers have suggested digital analysis could help reveal fraud in laboratory notebooks.

Inevitably, the increasing use of digital analysis will lead to greater awareness of its power by fraudsters. But according to Nigrini, that knowledge won't do them much good, apart from warning them off: 'The problem for fraudsters is that they have no idea what the whole picture looks like until all the data are in,' says Nigrini. 'Frauds usually involve just a part of a data set, but the fraudsters don't know how that set will be analysed: by quarter, say, or department, or by region. Ensuring the fraud always complies with Benford's law is going to be tough, and most fraudsters aren't rocket scientists.'

In any case, says Nigrini, there is more to Benford's law than tracking down fraudsters. Take data storage: mathematician Peter Schatte at the Bergakademie Technical University, Freiberg, has come up with rules that optimise computer-data storage, by allocating storage space according to the proportions dictated by Benford's law.

Ted Hill at Georgia Tech thinks that the ubiquity of Benford's law could also prove useful to those such as Treasury forecasters and demographers who need a simple 'reality check' for their mathematical models. 'Nigrini showed recently that the populations of the 3,000-plus counties in the US are very close to Benford's law,' says Hill. 'That suggests it could be a test for models which predict future populations: if the figures predicted are not close to Benford, then rethink the model.'

Both Nigrini and Hill stress that Benford's law is not a panacea for fraud-busters or the world's data-crunching ills. Deviations from the law's predictions can be caused by nothing more nefarious than people rounding numbers up or down, for example. And both accept that there is plenty of scope for making a hash of applying it to real-life situations: 'Every mathematical theorem or statistical test can be misused; that does not worry me,' says Hill.

But they share a sense that there are some really clever uses of Benford's law still waiting to be dreamt up. Says Hill: 'For me the law is a prime example of a mathematical idea which is a surprise to everyone, even the experts.'

Note: Alex is not the real name of Nigrini's former student.

Here, there and everywhere

Nature's preferences for certain numbers and sequences has long fascinated mathematicians. The so-called Golden Ratio, roughly equal to 1.62: 1 and supposedly giving the most aesthetically pleasing dimensions for rectangles, has been found lurking in all kinds of places, from seashells to knots. Then there's the Fibonacci sequence – 1, 1, 2, 3, 5, 8 and so on – where every figure is the sum of its two predecessors. This crops up all over the place in nature, from the arrangement of leaves on plants to the spiral pattern of seeds in the head of a sunflower.

Benford's law appears to be another fundamental feature of the mathematical universe, with the percentage of numbers starting with the digit D given by 100 times the base-10 logarithm of $1 + (1/D)$. In other words, around $100 \times \log_{10}(1 + 1/1) = 30$ per cent of such numbers will begin with '1'; $100 \times \log_{10}(1 + \frac{1}{2}) = 17.6$ per cent with '2', down to $100 \times \log_{10}(1 + 1/9) = 4.6$ per cent starting with '9'.

But the mathematics of Benford's law goes further, predicting the proportion of digits in the rest of the numbers as well. For example, the law predicts that '0' is the most likely second digit, accounting for around 12 per cent of all second digits, while 9 is the least likely, at 8.5 per cent.

Benford's law thus suggests that the most common non-random numbers are those starting with '10 ...', which should be almost 10 times more abundant than the least likely, which will be those starting '99 ...'

As one might expect, Benford's law predicts that the relative proportions of 1, 2, 3 and so on making up later digits of numbers become progressively more even, tending towards precisely 10 per cent for the least significant digit of every large number.

In a nice little twist, it turns out that the Fibonacci sequence, the Golden Ratio and Benford's law are all linked. The ratio of successive terms in a Fibonacci sequence tend towards the Golden Ratio, while the digits of all the numbers making up the Fibonacci sequence tend to conform to Benford's law.

Let's get lost

In a fitting close, it's time to make a plea: let's not eradicate all uncertainty from our lives. From GPS to book recommendations, technology is making everything precise and predictable, but that's not necessarily a good thing. Happiness may even depend on taking a chance. That's certainly what Catherine de Lange found.

It is pretty easy to go unnoticed as I follow my target along the busy high road, but when she turns into a residential side street, I start to worry. I slow down a bit, hang back and follow the woman from a safer distance.

Soon she turns and cuts through a large and rather lovely park, and even though I'm just minutes from my home, I'm surprised to find I've never been here before. By the time I regain the streets on the other side, the woman is nowhere to be seen and I am lost. I get out my smartphone and check the GPS for directions. 'Turn right onto Gascony Ave,' it reads, 'then look for someone who seems lonely and ask to walk with them for a while.' Here we go again.

Following random strangers to see where I end up is not the way I usually choose to spend my Saturday afternoons, but maybe it should be. With the rise of technologies designed to streamline our lives – from GPS devices to recommendation services – little need now be left to chance. But an emerging body of research suggests that chance is a vastly under-appreciated ingredient in human happiness. And apps called serendipity generators are encouraging us to buck the ultra-efficiency trend by putting some whimsy back into our lives. Can they help us overcome our inherent fear of uncertainty?

The rise of these new apps echoes a much earlier protest against the tyranny of modern efficiency. In the mid-19th century, the order brought about by the revolution in France gave rise to a cultural phenomenon known as flânerie. Dissatisfied with the urgency and alienation of the modern-day city, Parisian flâneurs hoped to encourage a certain kind of aimlessly enjoyable wandering in city life. A century later, cities became even more predictable as planners increasingly built them to conform to rigid grids, and maps became ubiquitous. Artists and activists once again resisted the orderly pragmatism, this time by using those maps to go nowhere in particular. For example, a collective known as the Fluxus artists created tongue-in-cheek instructions to 'step in every puddle in the city'.

The early internet wouldn't have been a target for the disaffected flâneur. When it took off in the 1990s it was mainly populated by people sharing things they liked with people they didn't know; it was a way to engage with people we wouldn't normally meet. In other words, it was a pretty good serendipity engine.

Then something changed. 'Coming out of the 20th

century and into the 21st the rhetoric changed to one of optimisation,' says Mark Shepard, an artist who designs serendipity apps. 'Making things more efficient has dominated the way we think about what tech should do for us – it's the idea of machine as humble servant that makes life easier.'

With that shift came the rise of recommender systems, algorithms that use your purchases, likes and browsing history, as well as those of other people, to work out what future purchases you might be interested in.

Every smartphone now has GPS to guide you to almost any destination. From choosing what to buy in the supermarket to finding your way without getting lost, the device in your pocket can make sure you'll never have to rely on chance again. We are optimised to within an inch of our lives.

As if on cue, apps have arrived that echo the flâneurs and get you lost on purpose. Many are a direct critique of the recommender systems they spoof. 'These always send you to the safer options, at the expense of the more interesting places,' says Ben Kirman, a computer scientist at the University of Lincoln, UK, who specialises in social games.

That's why Kirman created Getlostbot, an app that encourages users to break out of old routines and try different places. Download it, and it will silently monitor your Foursquare check-ins. When you become too predictable, always going to the same bar on a Friday night, for example, Getlostbot will send directions to one you've never tried before.

Over the past two years, a host of similar apps and

services have quietly proliferated. Apps such as Highlight, for example, connect you with nearby strangers. An online service called Graze sends you boxes of surprise food.

The serendipity stunts of flâneurs and artists might have seemed purely whimsical, but recent findings from happiness research suggest they were zeroing in on a surprisingly profound conflict in human nature. Part of what makes recommender systems so appealing is that most of the time, removing uncertainty is a really good idea. 'Human beings are constantly trying to make sense of the world,' says Tim Wilson, a psychologist at the University of Virginia. Understand something, and you are better placed to make sure it happens again if it's good, or be able to prevent it if it's bad.

And so, when you're looking at the possibility of a bad outcome – be it a bad movie or getting hopelessly lost – nothing will make you more unhappy than uncertainty. Getting lost or being unhappy with a purchase is not life-threatening, but our reluctance to deal with uncertainty might be easier to understand in the context of its effect in far more serious situations. Consider, for example, a study of people who were waiting to discover the results of a genetic test for Huntington's disease. Those who found out their results – whether they were positive or negative – experienced a boost in well-being. But the story was different for the people whose tests were inconclusive: this group felt greater distress over the next year than even the people who had found out they would spend their lives with a life-threatening and debilitating disease.

Why is that? Numerous studies confirm that when

something unexpected happens, we respond more emotionally to it. The mechanism is the same whether it's amplifying a modestly unpleasant event or a very serious one: we spend longer thinking about it, trying to find an explanation. Once we come up with a reason, however, we adapt to it, integrating it into the mundane.

Taking the uncertainty out of life, then, seems like a good strategy for happiness.

Unfortunately, however, this picture is incomplete. Most research on uncertainty has tended to focus on the negative aspects, but over the past decade psychologists have begun to investigate its effect on good experiences. Their findings are building a strong case that the same mechanism that causes uncertainty to intensify bad scenarios could make it a crucial ingredient in happiness.

For example, Wilson had a theory that for pleasurable events, keeping the uncertainty would be beneficial. To test the idea he devised a series of experiments. In one study, participants were told they were being given the chance to enter a competition, and asked to choose the two prizes they would most like to win. All were then told they had won. One group received their favourite prize straight away. The other group, however, would not find out which of their two favourites they would receive until the study ended. Those who were forced to spend time mulling over the two possible happy outcomes, Wilson found, hung on to their good mood far longer than those who experienced instant gratification.

They also spent longer looking at pictures of their possible bounty, lending support to the theory that people spend more time fixating on possible outcomes when something is uncertain. For happy outcomes, that amplifies the pleasure that can be derived from them.

An ambiguous pleasurable event is by its very nature harder to make sense of, forcing you to focus on it for longer, prolonging your emotional high. This gives rise to a phenomenon psychologists call the pleasure paradox: we want to understand the world, but that understanding can rob us of the pleasure we get from unexpected events.

These findings are just a small part of a body of research revealing quite how much pleasure can be gained through the power of uncertainty, and suggesting that technologies that introduce an element of chance into our lives could boost our mood in the day-to-day.

This is why I find myself stalking a complete stranger through north London on a rainy afternoon. I am testing Serendipitor, a satellite navigation app that augments your directions with small suggestions that introduce minor slippages, detours or distractions.

Designers of such apps straddle a thin line between convincing people to take a risk and invoking the ire of people who think such apps are absurd. 'Serendipitor was an ironic approach to saying: what does it mean when we're living in a society in which we need to download an application for serendipity?' says Shepard, who designed the app. Unlike simple chance, however, serendipity-generating apps weight the dice to ensure a positive outcome. Graze, for example, lets you opt out of foods you genuinely hate. Serendipitor gets you lost even as you are plugged into the reliability of Google maps.

Having made lunch plans, I use the app to look up the route. The walk to the restaurant should take just 6 minutes, and my phone shows a predictable route down the main road. As soon as I set off, though, Serendipitor sets me my first challenge: pick a person to follow for two blocks (Shepard says he borrowed many of his off-beat instructions from the Fluxus artists). Singling out a woman with a wheelie suitcase, I fall in behind, and before long she crosses the road and leads me to the park I had never known was there. The advantages of the app are now starting to become clear, and I can't stop thinking about the fact that had I picked anyone else, I would have remained ignorant of this place.

I'm not the only one who is bewitched by thoughts of what might never have been. In 2008, Harvard University psychologist Daniel Gilbert recruited a group of people who were in happy relationships of at least five years. They split the group, asking half to write down the story of how they met their partner, and the others to describe ways the couple might have failed to meet. When quizzed afterwards, those who wrote about how they might not have met their partner were in a better mood – and felt a bigger boost in satisfaction with their relationship – than the group who wrote the true love story.

Wilson refers to this as the 'George Bailey effect' after the protagonist of *It's A Wonderful Life* who is shown a world in which he was never born. Thinking about all the ways a good thing might never have happened, he says, breathes new life into feelings that have long since lost the shine of novelty.

On top of the buzz I get from my chance encounter, it's also strangely exhilarating to be told to carry out random activities. After following the woman to the park, it took me a while to pluck up the courage to ask someone whether I could take their photo, but doing it left me with a pronounced, if silly, sense of achievement. Yet I can't help wondering, if I weren't on an assignment, would I actually use such an app?

The truth is, people consistently underestimate the positive effects of uncertainty. Nobody knows this better than Kirman, who has found that although people respond well to the idea of Getlostbot, when it does suddenly appear on the screen telling them they need to try something new, they are reluctant to do so. In other words, people love the app, download it, and then don't use it.

If our resistance to uncertainty isn't enough of a problem, another stumbling block to widespread adoption of serendipity is commercial. There's no money to be made from an app that gets you lost.

But that doesn't mean we don't need them. Our increasing reliance on recommendations means people can end up living in a 'filter bubble' that narrows their field of vision, according to danah boyd, at Microsoft Research in Cambridge, Massachusetts. She sums up the current approach to online technology as a mix of fear of the unknown and pressure to stay within these safe bubbles.

For this reason, boyd reckons that these technologies will never make it big in the mainstream, but she still thinks they represent a useful countervailing mindset. 'We've lost the recognition that connecting to people whose worldviews are fundamentally different is important.'

She traces that to a shift in attitude around 2005 when,

she says, media focus on online predators led to 'a moral panic around stranger danger'. Around the same time we saw a rise of social networks and people using the internet to connect only to familiar faces rather than people they did not know.

It's not just our online lives that have become limited. 'One of the most important things is letting your kids embrace serendipity,' says boyd. 'That's what it used to mean to get on your bike and go out to wherever. We have lost that.' Could this umbilical access to recommender apps, GPS and other safe technologies be changing people's tolerance for risk? Over the past few years, the Pew Research Center in Washington DC has found, among other things, that fewer US teens are learning to drive, the sale of bicycles has plummeted, and young people are less prepared to move to another state, even if it means a better job.

But there might yet be hope for engineered serendipity. Big companies have begun to play with the idea. In 2008, Apple reportedly applied for a patent on a system that automatically connects two devices if they stumble into close proximity – for instance, if you happen to find yourself in the same area as a friend without realising it. Google's Latitude application does the same thing.

I don't expect Google maps to start instructing me to follow strangers, but might the company use the technology to add, to its current 'fastest' and 'shortest' options, an option for 'most adventurous'?

After all, by injecting a little more surprise into the technology we use every day, we might start to notice again what we miss in our relentless quest for efficiency.

'This is the core storyline of the most popular books,' says boyd. 'They stumbled upon something random and it was magical, and off they went to the wilderness. We fantasise about these things, but how do we allow fantasy back into our reality?'

Acknowledgements

I am something of a chancer, but not enough to claim this as my own idea. The original notion of a book on chance and randomness can be traced back to *New Scientist*'s then editor-in-chief (now editor at large) Jeremy Webb. It had been rumbling around in the back of his mind for a good while before he suggested it to Andrew Franklin at Profile Books. Andrew, as shrewd as he is dynamic, jumped on the idea and set the wheels of publishing in motion.

Jeremy didn't just come up with the idea; he also completed the first trawl of *New Scientist*'s extensive archive for articles that would fit the brief. This book is the result of some iterations on his initial discoveries, but not terribly many, if I'm being honest. The gallant and self-effacing Jeremy should by rights be alongside me as co-editor, but he's far too humble and unassuming for that. It was all I could do to let me buy him lunch. If you enjoyed this book, and you should happen to see him, do say thanks.

To acknowledge Jeremy and Andrew is not enough. This book also owes its existence to *New Scientist*'s publisher, John MacFarlane, and a huge cohort of the magazine's highly skilled staff and writers. I am especially grateful to the contributing authors, who took the time to check my edits of their work, and suggest many improvements. I should also like to thank the *New Scientist* feature editors, subeditors and graphics staff whose hard work honed and polished the articles on which this book is built. Particular thanks are due to Richard Webb, who helped find and fill the random gaps in the *New Scientist* archive's provision. I must also recognise my debt to Paul Forty's skill: his management of this project turned a daunting prospect into a viable proposition and – in the end – one of those unceasingly beautiful things: a published book.

Traditionally, all mistakes are the editor's fault and responsibility. However, if you've read this far you'll understand that the rules need changing. In a universe built on quantum uncertainty, skewed by chaos and besieged by Bayesians, how can we ever know for sure?

(I did say I was something of a chancer.)

Michael Brooks

About the contributors

Anil Ananthaswamy ('I, algorithm', published 26 January 2011) is a former software engineer, a consultant for *New Scientist* and the author of *The Edge of Physics*. His latest book is *The Man Who Wasn't There: Investigations into the Strange New Science of the Self*.

Stephen Battersby ('Cosmic lottery', published 22 September 2010) is a freelance science writer, quiz-question setter and a consultant for *New Scientist*. He writes on everything and anything contained in this universe or any other.

Mark Buchanan ('God plays dice – and for good reason', published 22 August 1998) is an American-born physicist and author now living in Europe. He has worked as a feature editor with *New Scientist*, and writes columns for *Nature Physics* and *Bloomberg*. Mark is a co-founder of the media training consultancy Write About Science, and his latest book is *Forecast*.

Gregory Chaitin ('Known unknowns', published 24 March 1990) is a mathematician and computer scientist who worked for many years at the IBM Watson Research Center in New York and is currently a professor at the Federal University of Rio de Janeiro and an honorary professor at the University of Buenos Aires. He is the author of a dozen books about mathematics and philosophy, including *Meta Math! The Quest for Omega*, and *Proving Darwin: Making Biology Mathematical*. Many of his essays can be obtained from his website at https://ufrj.academia.edu/GregoryChaitin.

Jack Cohen ('That's amazing – isn't it?', published 17 January 1998) is a retired biologist and the author of numerous books including *What Does a Martian Look Like?* and *The Science of Discworld*.

Paul Davies ('The algorithm of life', published 18 September 1999; 'Your random future', published 6 October 1990) is director of the Beyond Center for Fundamental Concepts in Science at Arizona State University in Tempe.

Dylan Evans ('The arbitrary ape', published 22 August 1998) is a writer, researcher and entrepreneur. His latest book, *The Utopia Experiment* (Picador, 2015), describes his attempt at post-apocalyptic living in the Scottish Highlands.

Bob Holmes ('The accident of species', published 10 March 2010; The prepared mind', published 22 August 2015; 'A chance at life', published 13 March 2015) has been a correspondent for *New Scientist* for more than two decades, and has written in excess of 800 articles for the magazine. He has a PhD in evolutionary biology from the University of Arizona, and his book *Flavour: a user's guide to our most neglected sense* will be published in 2016.

Nick Lane ('A miraculous merger', published 25 June 2012) is a reader in evolutionary biochemistry at University College London. His latest book is *The Vital Question: Why Is Life The Way It Is?*

Catherine de Lange ('Let's get lost', published 30 August 2012) is Biomedical Features Editor at *New Scientist*.

Graham Lawton ('Asteroids with a silver lining', published 22 September 2010) is Deputy Editor at *New Scientist*.

Robert **Matthews** ('Is that supposed to happen?', published 25 September 2004; 'The power of one', published 10 July 1999) is a visiting professor in quantitative research at Aston University, Birmingham and a science writer based in Oxfordshire.

Henry Nicholls ('Bullet-proof', published 10 January 2011) is a journalist, author and broadcaster, specialising in evolutionary biology, conservation and history of science. His books include *Lonesome George: The Life and Loves of a Conservation Icon*, *The Way of the Panda: The Curious History of China's Political Animal* and *The Galapagos: A Natural History*. He writes the *Animal Magic* blog for *The Guardian* and is currently working on a book on sleep and sleep disorders.

Regina Nuzzo ('The probability peace talks', published 11 March 2015) started her own mathematics tutoring business at age 13 and has never looked back. She is now a professor teaching statistics at Gallaudet University in Washington DC. Her writing has appeared in numerous publications beside *New Scientist*, including *Nature*, *Scientific American* and the *Los Angeles Times*.

Kate Ravilious ('Locating, locating, locating', published 22 November 2006) is a science journalist based in York, UK, who also runs media training for science professionals. She has a particular passion for earth sciences and archaeology.

Angela Saini ('Rough justice', published 21 October 2009) is an award-winning London-based journalist who regularly presents science-themed radio shows for the BBC. Her latest book is *Geek Nation: How Indian Science Is Taking Over The World*.

David Shiga ('Cosmic lottery', published 28 September 2010) is a software engineer at the Broad Institute, a biomedical research organisation. In a previous existence he worked as a reporter for *New Scientist*.

Laura Spinney ('Do dither', published 18 June 2008) is a novelist, non-fiction author and science journalist. Her third book, *Rue Centrale*, is a study of life in a central European city. She is based in Paris, France.

Ian Stewart ('That's amazing – isn't it?', published 17 January 1998; 'In the lap of the gods', published 25 September 2004) is emeritus professor of mathematics at Warwick University. He has published over 80 books including *Professor Stewart's Incredible Numbers*, *Why Beauty is Truth* and *The Science of Discworld*.

Helen Thomson ('The chips are down', published 11 August 2009) spent eight years as an editor and reporter at *New Scientist*. Her first book, on the lives of people with the world's strangest brains, will be published in 2016.

Vlatko Vedral ('Who's in charge here?', published 24 November 2006) is a professor of physics at the University of Oxford and the Centre for Quantum Technologies at the National University of Singapore. His latest book is *Decoding Reality: The Universe as Quantum Information*.

Clare Wilson ('Lucky you!', published 6 June 2012) is a *New Scientist* writer and editor focused on medical reporting.

Richard Wiseman ('The luck factor') is a professor in the Public Understanding of Psychology at the University of Hertfordshire, where he researches the psychology of luck, change, perception and deception. He has written several bestselling books including *The Luck Factor*, *Quirkology*, and *59 Seconds*.

Index

Figures in *italics* indicate illustrations.